Numbers at Work

Numbers at Work

A Cultural Perspective

Rudolf Taschner

TRANSLATED BY OTMAR BINDER
AND DAVID SINCLAIR-JONES

A K Peters, Ltd.
Wellesley, Massachusetts

Editorial, Sales, and Customer Service Office

A K Peters, Ltd.
888 Worcester Street, Suite 230
Wellesley, MA 02482
www.akpeters.com

Originally published in the German language by Friedr. Vieweg & Sohn Verlag, 65189 Wiesbaden, Germany, as *Der Zahlen gigantische Schatten* by Rudolf Taschner (third edition). Copyright © Friedr. Vieweg & Sohn Verlag | GWV Fachverlage GmbH, Wiesbaden 2005.

Library of Congress Cataloging-in-Publication Data

Taschner, Rudolf J. (Rudolf Josef)
  [Der Zahlen gigantische Schatten. English]
  Numbers at work : a cultural perspective / Rudolf Taschner ; translated by Otmar Binder and David Sinclair-Jones.
      p. cm.
  Includes bibliographical references and index.
  ISBN-13: 978-1-56881-290-8 (alk. paper)
  ISBN-10: 1-56881-290-6 (alk. paper)
  1. Number concept. 2. Mathematics--History. I. Title.
  QA141.15.T27 2007
  510.9--dc22
                            2006025221

Printed in India
11 10 09 08 07   10 9 8 7 6 5 4 3 2 1

The union of the mathematician with the poet, fervor with measure,
passion with correctness, this surely is the ideal.

—WILLIAM JAMES, *COLLECTED ESSAYS*

Table of Contents

Preface

Albert Einstein used to talk about the unique thrill he experienced when he looked over the shoulders of the *Old One*—his metaphor for the Creator, even though he had serious doubts about the existence of a personal God—and caught a glimpse of his playing cards.

But what did he mean by this? Ever since Pythagoras mathematicians have been certain that God's "playing cards"—the building blocks of creation—are nothing other than numbers. This conviction was perhaps expressed most eloquently by Galileo:

> The whole of Philosophy is written in this grand book, the universe, which stands continually open to our gaze. But the book cannot be understood unless one first learns to understand the language and to decipher the characters in which it is expressed. This language is mathematics, and its characters are triangles, circles, and other geometric figures. Without a knowledge of these it is humanly impossible to understand a single word of this book and we are condemned to traipse around aimlessly, lost in a dark labyrinth.

It is true that Galileo (and later Kant) saw mathematics as founded in geometry, which is itself directly grounded in what comes to us via our senses. This is still a hair's breadth away from an actual peek into the cards of the Old One. Geometry does deprive sensory data of their color, corporealness, transience, vulnerability—in short, of all their opaque and baroque charms. Using straight lines, circles and their points of intersection, geometry attempts a representation of the bare bones of what we experience—what is left of our sensory data after abstraction has reduced it as far as it can. In geometry, the painter's inspired brush strokes are reduced to the transparent, limpid constructions of an engineer wielding compass and ruler. Yet, geometry always retains an irreducible residue that affects our senses. Numbers do not.

In the case of arithmetic, the theory of numbers, there is no direct link connecting it to the experience of our senses. While we can see and touch the silver pieces we are counting and mentally connect them to the number thirty, we cannot see or touch the number thirty itself. We hear the steeple clock strike, and we count up to eight—yet all that we hear is those strokes of the clock that we mentally subsume under eight; we do not hear the number eight itself. There is no way in which we can have direct sensuous experience of numbers, no way of experiencing them in their own right through any of our senses; neither our eyes nor our sense of touch nor our ears can give us access to them; they remain forever devoid of scent and taste.

We now know—and have known since David Hilbert's insights into the foundations of geometry at the latest—that all of geometry can be incorporated into the realm of numbers. All the insights and conclusions of geometry can be deduced solely from the laws of arithmetic without recourse to sight or touch.

And it is not only geometry: this dictum may very well apply to all of intelligible reality. While this is not what this book purports to prove, an attempt is made in it to *point* towards the extent to which numbers are at work in a great variety of aspects of reality.

It is significant for the essential paradox of the human condition that it is numbers—cut off from sensuous experience by definition—that grant us our deepest insights into the nature of reality. Perhaps this is best illustrated by an anecdote that once more has Albert Einstein at its center.

As a young nobody Einstein loved to go the zoo at Berne to watch the bear-feeding ritual. He noticed that the bears mostly trotted around with their snouts close to the ground, so that they only found those goodies they bumped into. However, sometimes one of them would rear itself up on its hind legs to gain a vantage point from which it could spot the choicest morsels. This reminded Einstein irresistibly of the majority of physicists, seeing no further than their noses as they crouch over their calculations. Truly significant discoveries are only made by those who survey the broad context. It is only by applying abstraction till it yields the numbers at the heart of things that we may hope to raise ourselves to such heights that we may sneak a peek at God's playing cards.

RUDOLF TASCHNER

Translators' Note

Translating Professor Taschner's work proved a far more time-consuming labor than either of us had imagined when we first agreed to undertake this project. It was not the difficulty of rendering thoughts from German to English—we were familiar enough with that process and had allowed sufficient time for it. Still less was it any difficulty in liaison or interpretation. The problem was the work itself, and *problem* is perhaps not the word that best describes what delayed our arrival at the finishing post.

Reading *Numbers at Work* is like walking down a great corridor lined with books: there is much to see on the shelves, much to learn from each individual book, and great benefit to be gained from reaching the end. The trouble for the translator, and for all subsequent readers, is that the corridor is studded with doors, each of them leading into a corridor of its own, with shelves and books and doors of its own. It is very difficult to press ahead without investigating at least some of these distractions. We strongly advise you not to try—to give in to temptation and to follow each tantalizing lead: there is, as Galileo pointed out, a whole universe to explore.

OTMAR BINDER

DAVID SINCLAIR-JONES

Pythagoras

Numbers and Symbol

Numbers are the key to all knowledge, according to Pythagoras of Samos. He believed that they underlie the whole cosmos and all that is in it, our very selves and all that goes to make us what we are. Nothing exists that cannot be expressed as numbers.

Nobody knows how Pythagoras arrived at this idea. He wrote nothing—or at least nothing that has come down to us. Nonetheless, there is enough in the historical record to allow us some room for conjecture.

1 Pythagoras of Samos.

2 Thales of Miletus.

3 His knowledge of Babylonian astronomy enabled Thales to predict the solar eclipse of 585 BC.

Thales of Miletus, Pythagoras' elder by less than a generation, lived around 600 BC. Pythagoras traveled widely, always seeking out experiences and people who could help him in his studies. It is very possible that he met Thales, and it is almost certain that he was influenced by the man who is still seen as the prototypical philosopher.

Thales' teaching career can be dated with reasonable precision. He is credited with predicting the solar eclipse of May 28, 585 BC. This feat was stunning enough as a purely scientific achievement. What is said to have followed is surely one of the most important developments in the intellectual history of mankind.

At the precise moment of this eclipse, a battle was raging between armies led by Alyattes of Lydia and Cyaxares of the Medes. The Medes, shocked by the sudden switch from day to night, dropped their arms and fled, leaving the victory to Alyattes, who held his ground. Alyattes had been forewarned of the celestial phenomenon by Thales.

We may doubt the historical authenticity of this account,[1] but the message it conveys is unique as a turning point in the intellectual history of mankind. Rational thinking applied to experience enabled Thales to predict the eclipse; foreknowledge of the eclipse helped Alyattes to victory. The losers, hobbled by ignorance, put the eclipse down to the wrath of the gods and fled. The victors, knowing it to be a natural phenomenon brought about by an understandable sequence of events, held their ground and won. It was not the whim of the gods, creatures of myth, that helped Alyattes to victory; rather, it was the conclusions of Thales, drawn solely from what the Greeks called *lógos*.[2]

Thales, then, is a likely source for this key insight in Pythagoras' thought: ours is not the role of inert playthings at the mercy of irrational, unfathomable divine forces. The universe is not chaos running riot; rather, we are part of an orderly cosmos that awaits our rational interpretation.

Perhaps[3] it was at this point that Pythagoras raised an obvious, yet at the same time, all-decisive question: What is the basis of understanding? What are, as it were, its basic building blocks, its atoms? Where does it all begin? What is so simple and so self-evident that no further explanation is needed? What are the axioms it would be pointless to question, because everything about them has been established beyond doubt?

Pythagoras came to believe that he had an answer to that question. Nothing, as he saw it, was more elemental than counting. For someone who had grasped the principle of counting, beginning with *one* and working up to ever-new numbers by continually adding *one*, it was simply inconceivable to proceed in any other way. Counting is an activity in which all humanity, its manifold differences in all other areas notwithstanding, behaves as one on a global scale. In fact, it is generally agreed today that if radio contact with intelligent extraterrestrials were possible,[4] this contact would have to be couched in the terms of counting. Counting, in effect, is seen as the only technique common to all intelligent beings.

Here, then, is the *axiom*: only when we understand what numbers are at the core of a certain state of affairs can we be certain that we actually understand it. To understand something completely means to understand it as self-evidently as we understand counting.

Thales leads us to believe that we can understand the world. It follows, for Pythagoras, that the world must exist as numbers, because it is only when a state of affairs is reduced to numbers that it can be understood.

One may object that there are civilizations that are not acquainted with the idea of numbers. Some peoples—for instance, the *Abipones* and the *Yonoama* in Latin America or the *Rumilara* in Southern Australia—are effectively innumerate, capable of only grasping ideas of entities in the singular, in pairs and, at the very most, in triads. Confronted with more than three items, Brazilian *Bakairi* or *Bororo* Indians simply see a *multitude* and grab their hair to express that diagnosis. For them, the question "How many precisely?" is simply not one that comes to mind.

We can furthermore object that we simply do not know how our understanding of numbers has come about. After all, is not the discovery of numbers hidden in the dawn of history?

To begin with, numbers were tied inseparably to the objects to be counted. In early Mesopotamian times, two different *fours* were needed for *four bushels of wheat* and *four beef carcasses*. Vestiges of this proto-understanding of numbers have been preserved until today. Having tried on *two* shoes, we buy the *pair*—we make a distinction between *pair*, which

4 Early Sumerian numerical and script characters.

denotes two-ness at a stage that precedes counting, and *two*, a full-fledged number. Number as an idea, number emancipated from the type of object being counted, seems to have emerged in Mesopotamia as early as the ninth century BC. If a trader took to the road with goods—say, five heads of cattle and seven sheep—he would have carried a box containing a representation of these goods, perhaps five clay balls and seven tablets. To be on the safe side, the contents of the box may have been displayed in pictures on the outside—this at least is how we interpret certain archaeological finds dating from the period. The next step was the realization that the pictures themselves provided sufficient information: a big step on the road toward script characters. Early Sumerian numerical symbols and script characters indeed show a close resemblance to these pictures. The ability to count seems to be inextricably intertwined with the skills of verbalizing, writing and reading.

Leaving these objections aside, substantial though they may be, it is clear that Pythagoras became convinced he could *prove* that numbers are the basic building blocks of everything that exists. He had discovered that dividing a single string (on an instrument called the *monochord*) into ratios of small whole numbers, such as 1:2, 2:3, 3:4, etc., generated musical intervals. These intervals, when used melodically, were capable of moving listeners to tears and speaking directly to their souls. This insight is laid down in the wonderful words of one of his pupils, Philolaus: "The soul is the numerical harmony of the body, and the soul's relationship to the body is the same as the one existing between notes and the musical instrument that produces them."

5 The wall of the "geometry room", Nicolaus Copernicus' study in Cracow, is full of sketches of the planets with their relative orbits.

Even celestial bodies were seen as restricted in their movements to spheres whose relationships, according to Pythagoras, were based on numerical ratios. Legend has it that these movements so enthralled Thales that his enthusiastic study of them once led to his falling into a well. These ratios of small whole numbers produced chords of a beauty that was held to be quite literally supernatural in that they could only be perceived by the gods: what has been called the *music of the spheres*.

The idea of such planetary spheres survived into the modern age. As late as 1596, a 25-year-old surveyor by the name of Johannes Kepler published his *Mysterium cosmographicum* (*The Secret of the Universe [Explained]*) in Graz, then the capital of the Habsburgs' southern possessions. In this book, he accounts for the distances between the sun and the planets by postulating a mutual interpenetration of the five regular Platonic bodies. These bodies determine the radii of the calottes or spherical surfaces that support the planets. By 1609, in his *Astronomica nova* (*The New Astronomy*), Kepler himself had discarded this concept, one of the legacies of antiquity, and cleared the way for the modern explanation of celestial mechanics.

But let us return once more to antiquity. It is not surprising that Pythagoras, in the euphoria of a great discovery, drew a great number of inferences from it, the majority of which were misconceived (the idea of the music of the spheres being one example) and led only to further errors. Certainly the inspired number symbolism of the Pythagoreans, like the numerologies of other cultures, stayed at surface level and did not penetrate to a deeper understanding. Before we try to explain why this was so, let us spend a little time in the company of those who have tried to seek knowledge through the symbolic power of numbers.

Following age-old tradition, Pythagoras attempted to identify geometrical patterns related to numbers. He meditated on triangular numbers

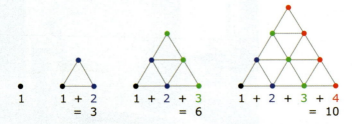

and so on, on square numbers

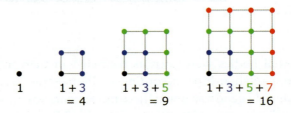

and so on, and on similar numerical patterns. He differentiated between *even* numbers, which can be arranged in half as many pairs, and *odd* numbers, the pairing of which always leaves a remainder of one. Interestingly enough, for some advocates of numerical symbolism, odd numbers are considered to be "good"; even numbers, "evil". This may be because if odd numbers are added one to another in sequence beginning with *one*, the answers give the square numbers in sequence: $1 + 3 = 4$, $1 + 3 + 5 = 9$, $1 + 3 + 5 + 7 = 16$, and so on.[5] No such relationship exists for even numbers.

Geometrical patterns such as those mentioned earlier were seminal for numerical symbolism. It is interesting to note that even in antiquity, *one* had a special status among numbers: it was regarded as the unit and did not normally figure in counting. This is why *one* was used as the symbol of the indivisible and, ultimately, of the divine.

For the ancient Greeks, *two* was the first genuine number. It symbolized opposition, antithesis, disunion: the polarity of man and woman, of left and right, of good and evil, active and passive, sun and moon, day and night.

6 Yin and Yang.

In the Christian tradition, this motif is taken up by the polarity of the Old and the New Testament as well as that of man and God—foreshadowing Jesus, who, according to Christian theology, unites these two natures—human and divine—in himself.

Judaic tradition shows its own familiarity with the symbolic significance of the number two, which can most clearly be seen in the two tablets that Moses received for his people from God.

Three, the typical triangular number (and the first after *one*), symbolizes a kind of completeness in its geometrical pattern: the area of a triangle, which is defined by reference to its three vertices. This completeness is closely related to the idea of perfection. The trinity of mother, father and child has been seen from ancient times as a blueprint for human social life. In antiquity, this trinity often took the form of divine triads: Anu, Enlil, Ea in ancient Babylon; Brahma, Vishnu,

7 Trinity of mother, father and child.

Shiva in India; and, of course, the Christian trinity of Father, Son and Holy Spirit. In the Old Testament, the number *three* is of symbolic significance in several passages. In Isaiah, for instance, we find the following, in the words of the *Authorized Version*:

> In the year that king Uzziah died, I saw also the Lord sitting upon a throne, high and lifted up, and his train filled the temple. Above it stood the seraphims: each one had six wings; with twain he covered his face, and with twain he covered his feet, and with twain he did fly. And one cried unto another, and said, Holy, holy, holy is the Lord of hosts: the whole earth is full of his glory.

In Genesis, Abraham is visited by God—the one and only God—in the shape of three men. This happens in a mysterious manner, which the text expresses by shifting from three to one:

> And the Lord appeared unto him in the plains of Mamre; and he sat in the tent door in the heat of the day;
>
> And he lifted up his eyes and looked, and, lo, three men stood by him; and when he saw them, he ran to meet them from the tent door, and bowed himself to the ground,
>
> And said, My Lord, if now I have found favour in thy sight, pass not away, I pray thee, from thy servant....

And Genesis describes the creation of man thus:

> So God created man in his own image, in the image of God created he him, male and female created he them.

The perfection of creation is emphasized by the threefold repetition of "created" in three lines.

Four is the typical square number (again, the first such number after *one*). It symbolizes cosmological constructs: the four points of the compass, the four wind directions, the four seasons and Empedocles' four elements of fire, water, earth and air. The Bible gives us the four rivers Pishon, Gihon, Tigris and Euphrates that encircle the Garden of Eden and Daniel seeing in his dream four beasts (symbolizing the rulers of the global empire) emerging from a sea lashed by four winds.

The symbolic significance of the numbers *three* and *four* is further underlined in the Bible by the fact that the Israelites had three patriarchs—Abraham, Isaac and Jacob—and four archetypal mothers—Sarah, Rebecca, Leah and Rachel.

8 Abraham is visited by God in the guise of three men.

9 Compass rose.

Seven (the sum of *three* and *four*) and *twelve* (their product) are themselves of immense numerological importance.

Seven stands as the number of planets (in the ancient sense of heavenly bodies that change their position with reference to the fixed stars) visible to the naked eye: Sun, Moon, Mercury, Venus, Mars, Jupiter and Saturn. In this way, it evokes the perfection of the universe. The Old Testament is replete with references to the number *seven*, and Christianity has continued this tradition. Consider the seven loaves of bread with which Jesus feeds the four thousand, and the seven baskets full of scraps left over (Matt. 15:34–37).[6] Consider the seven petitions in the "Our Father", the seven sacraments, the seven virtues, the seven deadly sins, the seven gifts of the Holy Spirit and the seven works of charity.

Twelve is the number of the zodiac in the nocturnal sky of the Northern Hemisphere; the ancient Egyptians subdivided the day into twelve hours and the year into twelve months; the Gnostics proclaim twelve eons. There are twelve tribes of the Children of Israel; the Church as the New Israel acknowledges twelve apostles; in the Old Testament Book of Judges, the number of great judges is also twelve.

Doubling numbers strengthens their potency as symbols. In the Isaiah verses quoted earlier, the seraphs have six wings arranged in three pairs. The number *six* is the next triangular number after *three*. It is also the result of doubling *three*. Hence, its geometrical pattern is that of the *Magen David*, the Shield (or Star) of David, which consists of two equilateral triangles. Analogous to this is *eight*, whose geometrical pattern is the octagon, two superimposed squares, one of which has been rotated by 45°, each of them symbolizing the typical square number *four*. The octagon is a recurring feature in architectural design and in the arts. In Buddhism, a circle subdivided into eight segments, which derives from the octagon, is an emblem of the eight different paths on which to escape from the vale of tears of imperfect existence. Murray Gell-Mann, winner of the 1969 Nobel Prize for Physics, discovered a new and completely unexpected significance for *eight*: it represents the family of elementary particles to which protons and

10 In the ancient and medieval cosmos, the seven *planets*—the Moon, Mercury, Venus, the Sun, Mars, Jupiter and Saturn—and the twelve constellations go around the Earth.

11 The Shield of David.

12 The octagon in the dome of Aachen.

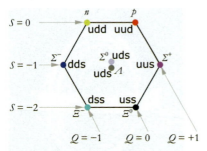

$S = 0$

$S = -1$

$S = -2$

$Q = -1$ $Q = 0$ $Q = +1$

13 The octet of heavy particles according to Gell-Mann.

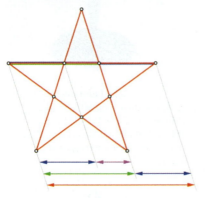

14 The diagonals of a regular pentagon form a pentagram. They intersect with one another so that the ratio between the entire length and the longer section is the same as that between the longer section and the medium one and between the medium section and the short one. This threefold ratio is the *Golden Mean*.

neutrons, the constituent components of the nucleus, belong.

Five generates the pentagram; for the Pythagoreans, this was the most mysterious geometrical pattern of them all. If you draw the five diagonals in a regular pentagon, a new regular pentagon appears in the middle, where again the diagonals can be drawn and so *ad infinitum*. The line sections produced by the intersecting diagonals form an aesthetically highly satisfying ratio, which later came to be called the *Golden Mean*. The Golden Mean appears in countless proportions in the fine arts and in architecture. In addition to this, the task of defining the Golden Mean by means of a numerical proportion proved an insoluble problem for Pythagoras' pupils. It was even insoluble for reasons of principle, as they noted to their astonishment, but this is discussed in greater detail later.

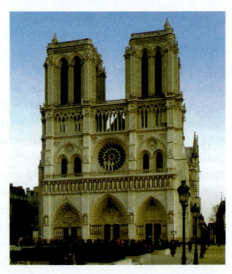

15 The cathedral of Notre Dame has many examples of the *Golden Mean*. Examples are the ratio between the total width and the width of one of the towers, the ratio between the height of the entrance story and that of the middle story with the rosette and the ratio between the height of the middle story to the top story.

Forty, in itself a number of no particular numerological interest, still has a very powerful symbolic significance in the broad context of purification, which survives to this day in the etymology of the word *quarantine*. We should also cite the belief that the fortieth day marks the end of an acute illness and the division of pregnancy into seven periods of forty days. *Forty* features prominently in Jewish and Islamic mysticism, in the Hebrew Bible and in the New Testament.[7] Some examples are the Israelites' forty years wandering through the wilderness (Num. 32:13), the duration of the Flood (Gen. 7:17), Moses' two sojourns on Mt. Sinai (Exod. 24:18) and Jesus fasting in the desert (Matt. 4:2).

Numbers that refused to be harnessed in geometrical patterns, or "refractory" numbers, were of particular interest to the Pythagoreans. The best known among these are the *prime numbers*. It is impossible to represent such numbers in rectangular patterns: 2, 3, 5, 7, 11, 13 or 17 dots can only be arranged in one line for rectangular presentation, whereas 9, 12 and 15 can be arranged in two or more equal lines.

The horizontal and the vertical extension of a rectangular number (i.e., the number of dots in the two directions) are obviously divisors of this number. The preceding figure shows that 9 has the divisor 3, 12 has the divisors 4 and 3 (and 6 and 2 as well), and 15 has the divisors 5 and 3. Prime numbers, on the other hand, have only one and themselves as possible divisors.[8] Even numbers always have two as a divisor (and are therefore never—with the exception of two itself—prime numbers).

For a number like 31 ($2 \cdot 3 \cdot 5 + 1$), 2, 3 or 5 are not possible divisors. If you try to express 31 as a rectangular pattern with one of these prime numbers as its horizontal extension, there will be a remainder of 1. Similarly, a number like 211 ($2 \cdot 3 \cdot 5 \cdot 7 + 1$) cannot have either 2, 3, 5 or 7 as divisors. Divisions by one of these prime numbers must again always leave a remainder of 1. Calculations like these led Euclid to conclude that no list of prime numbers, no enumeration of a *finite* number of prime numbers, was ever going to be exhaustive.[9]

Today we tend to say rather grandiosely, "There is an infinite number of prime numbers," but no equally grandiose proof is available. There is no magician to whisk a silk cloth off the table and reveal infinite cohorts of prime numbers happily disporting themselves!

In addition to geometrical patterns being generated by numbers, there is the attractive reverse procedure: arranging numbers in such a way that they can be fitted into geometrical patterns. Going back at least as far as Pythagoras' days, we find a number of different civilizations experimenting with "magic squares". A magic square consists of $3 \cdot 3 = 9$, $4 \cdot 4 = 16$ boxes or a square number of boxes arranged within a square. Each square has to be filled with the numbers 1, 2, 3, ..., n (where n denotes the number of boxes in the square) in such a way that every row, every column and the two diagonals add up to the same sum.

In China, a magic square made up of three rows and three columns called *Lo-Shu* is still sold today as a lucky charm. The sum in every row and every column and in the two diagonals must add up to 15, so we call 15 the *magic number* of any 3 × 3 square. This follows from the fact that the sum of the numbers 1 to 9 is 45. When this is divided into three equal addends, 45 yields 15, which therefore emerges as the *Lo-Shu* magic number. There are

4	9	2
3	5	7
8	1	6

16 Lo-Shu.

only a few ways to break down 15 as the sum of three numbers between 1 and 9; the following table gives a systematic survey:

$$9 + 5 + 1 = 15 \quad 8 + 6 + 1 = 15 \quad 8 + 4 + 3 = 15 \quad 7 + 5 + 3 = 15$$
$$9 + 4 + 2 = 15 \quad 8 + 5 + 2 = 15 \quad 7 + 6 + 2 = 15 \quad 6 + 5 + 4 = 15$$

All of these are realized in the *Lo-Shu*, this is why there are no 3 × 3 magic squares—with the exception of simple mirror inversions—found outside the *Lo-Shu*.

The magic number of a 2 × 2 square is $5 = (1 + 2 + 3 + 4)/2$. However, 5 can only be broken down into two addends ($4 + 1$ and $2 + 3$), which cannot provide us with a 2 × 2 magic square.

17 The magic square is above the angel's wing in Dürer's *Melencolia*.

The magic number of a 4 × 4 square can be found by the formula

$$1 + 2 + 3 + \dots + 14 + 15 + 16 = 136$$

and is therefore 34 (= 136/4). There are 86 possible ways to arrive at this sum using four addends between 1 and 16. It is not surprising, therefore, that there are many variations[10] on the 4 × 4 magic square. The most famous magic square is found in one of Albrecht Dürer's engravings, his 1514 work "Melencolia", which was occasioned by the death of the artist's mother in the same year. The artist has depicted a magic square in the upper right corner of the picture. Dürer constructed this magic square with the utmost ingenuity. He started with a square, in which the numbers 1 to 16 consecutively filled the boxes row by row.

1	2	3	4
5	6	7	8
9	10	11	12
13	14	15	16

In a second step, he made the pairs of numbers at the endpoints of the diagonals (1, 16), (4, 13) swap places and proceeded in the same way with those in the middle of the outer rows (2, 3), (14, 15) and with those in the middle of the two inner columns (6, 10), (7, 11). He did not move the two remaining pairs (5, 9) and (8, 12).

1	2	3	4
5	6	7	8
9	10	11	12
13	14	15	16

16	2	3	13
5	6	7	8
9	10	11	12
4	14	15	1

16	3	2	13
5	6	7	8
9	10	11	12
4	15	14	1

As expected, Dürer's square meets the requirements any magic square has to fulfill: each row, each column and each of the two diagonals add up to the magic number 34.

16	3	2	13
5	10	11	8
9	6	7	12
4	15	14	1

However, there is more to the ingenuity of Dürer's square. The four numbers 16, 13, 4, 1 in the corner boxes also add up to the magic sum of 34, as do the numbers 16, 3, 5, 10 in the boxes in the upper left of the square. The same sum, 34, is found for the numbers in the boxes in the upper right, those in the lower right and those in the lower left. Finally, the numbers 10, 11, 6, 7 in the boxes forming the center of the square also sum to the magic number 34.

Unbelievable as it may seem, this is only the beginning: there are 86 possibilities for four numbers between 1 and 16 to yield the magic number of 34 as their sum. In our imagination, we may connect all of the 86 sets of numbers that add up to 34 in Dürer's square so that we get a geometrical figure. It will then be possible to arrange in Dürer's square all of these 86 figures plus their respective mirror images as geometrically pleasing combinations.

This may appear amazing, yet there is more to come. A glance at the bottom line reveals a further dimension to the square's power of fascination. The two middle numbers form the number 1514, the year Dürer engraved the square, and the two numbers 4 and 1 in the two corners may be interpreted as signifying the fourth and the first letters of the alphabet, D and A, Albrecht Dürer's initials.

Here we see again the connection between numbers and script that dates back to those early times when numerical and script characters were being invented side by side by the first fully developed civilizations.

For the ancient Greeks, this connection was very much in evidence. In their script, the letters of their alphabet doubled as numbers. The first nine letters of the most ancient Greek alphabet also serve as symbols for the numbers one to nine; the remaining letters were used to denote bundles of tens and hundreds.[11] In a similar manner, the Jews used Hebrew script characters to symbolize numbers, א (aleph) for one, ב (bet) for two, ג (gimmel) for three, ד (dalet) for four, and so on. The equation of letters with numbers opens up an enormous speculative field for possible interpretations of sacred texts. Through the deliberate choice of letters in the composition of this text, the author may attempt to convey to those readers who have been initiated into number mysticism a deeper meaning than words alone can give.

Examples of this abound. In Genesis, Chapter 14, Abraham and his 318 servants are said to have pursued the captors of his nephew Lot. The 318 servants are actually reduced to one,

$$\text{אליעזר}$$

$$+ 7 + 10 + 1$$
$$318 = 200 + 70 + 30$$

18 "Eliezer" corresponds to 318.

Eliezer, because the Hebrew letters that make up his name signify 318 when interpreted as numerical symbols. A friend in need is like 318 indeed — what an effective way of saying how important Eliezer's help was to Abraham!

In Genesis, Chapter 28, Jacob dreams of a ladder "set up on the earth, and the top of it reached to heaven". This ladder, *sulam* in Hebrew, signifies Mount Sinai in the interpretation of some exegetes. The numerical values of the two words *sulam* and *Sinai* are the same; namely, 130. This follows a cogent logic: the laws revealed to Moses on Sinai are the ladder that connects heaven and earth.

19 Jacob's Ladder: "And Jacob went out from Beersheba and went toward Haran. And he lighted upon a certain place, and tarried there all night, because the sun was set; and he took of the stones of that place, and put them for his pillows, and lay down in that place to sleep. And he dreamed, and behold a ladder set up on the earth, and the top of it reached to heaven: and behold the angels of God ascending and descending on it. And, behold, the LORD stood above it, and said, I am the LORD God of Abraham thy father, and the God of Isaac" (Gen. 28:10–13).

The account of the creation of the world in Genesis begins with "In the beginning God created the heaven and the earth". The very first sentence of the Bible in Hebrew is *"bereishit bara Elohim et ha-shamayim ve-et ha-aretz"*.

If we were versed in numerology, we could turn that sentence into an acrostic[12] or acronym, consisting of the first letters of these seven[13] words, ב from *bereishit*, ב from *bara*, א from *Elohim*, etc. We could then add up the numerical values of the letters: $2 + 2 + 1 + 1 + 5 + 6 + 5 = 22$ and discover that the same holds true for the account of creation as a whole. Note that there are nine introductory phrases of "God said" and a tenth, slightly longer one of "God said to them" (three words in Hebrew). After the sixth "God said", there is the phrase, "And God blessed them, saying…." In Hebrew, the sum total of the words in these introductory phrases governed by "say" is 22. The fact that we encounter 22 twice in this context cannot be regarded as mere chance. The number 22 is, after all, the number of letters in the Hebrew alphabet, which gives rise to the following interpretation.

Creation is the embodiment of speech. Existence cannot be understood in terms of being thrown into a void; it has been called into being by the spoken word of the Creator. Language, of which letters with their double function of script and numerical characters are symbolic representatives, is therefore at the heart of Creation.

בראשית ברא אלהים
את השמים ואת הארץ

$$22 = \begin{matrix} & +\ 1 & +\ 2 & & +\ 2 \\ & 5 & +\ 6 & & +\ 5 + 1 \end{matrix}$$

20 The first letters of the first seven words of the Bible.

This is why we find the following passage in the kabalistic *Sefer Yesira*, or *The Book of Formation*, often attributed to Akiba ben Joseph:

> Twenty-two letters: he carved them out, he hewed them, he weighed them, he exchanged them, he combined them and formed with them the life of all creation and the life of all that would be formed. (Translated from the Hebrew by A. Peter Hayman.)

The unfolding of creation is mirrored in how words are formed from letters. Into how many possible combinations of two different letters can each of the 22 Hebrew letters enter? Each of them can be placed either before or after any of the others; the first letter, א, can be combined with the remaining 21 letters in this way, and the same holds for the second letter, ב, which can be combined with any of the remaining 20 in this way. The third, ג, can be combined with the remaining 19 and so on. To calculate all of these possibilities, this is the procedure:

$$(2 \cdot 21 + 2 \cdot 20 + 2 \cdot 19 + \ldots + 2 \cdot 3 + 2 \cdot 1) \text{ or, by reduction, } (22 \cdot 21).$$

The resulting sum is 462.

If one excises from the passage in which Genesis deals with creation the fundamental first sentence with its seven words, "In the beginning God created the heaven and the earth", which, as we have seen, programmatically refer in their acronym to the 22 individual letters of the alphabet, exactly 462 Hebrew words remain. It follows that the author of the account of creation not only wanted to give a graphic depiction of the sequence of events from the creation of light to the creation of the human race, he also wanted to explore, through the ingenious composition of his text, the unfolding of the alphabet in numerological-symbolist terms and the unfolding of numbers by means of words. Because Hebrew uses the same characters to depict letters and numbers, both objects can be achieved with the same text.

In this way, Moses describes in Genesis how the world came into being from the word of God. In the second book of Exodus (3:13–15), Moses is confronted with the burning bush and humbly asks the Lord to reveal his name. This scene is reproduced again in the pithy language of the *Authorized Version*:

> And Moses said unto God, Behold, when I come unto the children of Israel, and shall say unto them, The God of your fathers has sent me to you; and they shall say to me, What is his name? What shall I say unto them?

> And God said unto Moses, I AM THAT I AM: and he said, Thus shalt thou say unto the children of Israel, I AM hath sent me unto you.

And God said moreover unto Moses, Thus shalt thou say unto the children of Israel, THE LORD GOD of your fathers, the God of Abraham, the God of Isaac, and the God of Jacob, hath sent me unto you: this is my name for ever, and this is my memorial unto all generations.

In the burning-bush dialogue, we come across the *tetragrammaton* ("word with four letters"), which veils the ineffably holy name JHWH in secrecy. It is no wonder that its numerical value, 26 (10 + 5 + 6 + 5), is also concealed in the burning-bush dialogue as a whole. The dialogue consists of four utterances in direct speech. Moses' question draws three answers from his God. The words of Moses' question plus those of the second divine answer equal 26 words; similarly, the words of the first and the third divine answer also add up to 26. The revelation of the divinity's name is thus sealed twice with the number obtained from JHWH.

The Jewish God reveals himself through words, through structures consisting of the 22 Hebrew letters. This is why the last two clauses, "this is my name for ever, and this is my memorial unto all generations", consist of (10 + 12 =) 22 Hebrew letters. In addition to this, the whole burning-bush dialogue totals 253 letters. The sum of the numbers from 1 to 22 is 253; therefore, Moses thus conveys

21 Moses before God: "And when the LORD saw that he turned aside to see, God called unto him out of the midst of the bush, and said, Moses, Moses. And he said, Here am I. And He said, Draw not nigh hither: put off thy shoes from off thy feet, for the place whereon thou standest is holy ground" (Exod. 3:4–5).

to us that the sum of all the letters of the Hebrew alphabet, 253, is concealed in the burning-bush dialogue and the name of God is concealed in the sum total of the letters.

As in an echo from afar, we hear Pythagoras' claim, "Everything is numbers", in the words of the Bible.

The symbolic meaning of many of the numerical specifications to be found in the Jewish Bible, the Gospels and other venerable old texts can no longer be retrieved. For instance, why the patriarchs before the Flood lived to such ripe old ages as Moses describes can hardly be unraveled today. In the case of Enoch, however, who "walked with God" and lived to 365, the parallel to the days of the solar year is obvious. The patriarchs of the tribe of Israel also had lifespans that are interesting from a numerological point of view: Abraham died at $175 = 7 \cdot 5 \cdot 5$, his son Isaac at $180 = 5 \cdot 6 \cdot 6$ and Isaac's son Jacob at $147 = 3 \cdot 7 \cdot 7$. How are these numbers to be interpreted? Does the fact that the sum of the numbers in each formula is the same, $7 + 5 + 5 = 5 + 6 + 6 = 3 + 7 + 7 = 17$, point to some hidden message?

Matthew, as is generally known, begins his gospel with the three times 14 generations from Abraham to Jesus. In this, he may have been inspired by the name *David*, whose letters symbolize 14 ($4 + 6 + 4$) in Hebrew. Possibly, Matthew wants to convey that Jesus is truly the Messiah, being, as it were, David three times over, but this interpretation is by no means assured.

In the last chapter of the Gospel of St John, we read the story of Peter's miraculous draft of fish, when he caught 153 fish with no damage to his nets. What is the meaning of 153? Is the author referring to one of the following curious characteristics of 153?

$$153 = (1 \cdot 1 \cdot 1) + (5 \cdot 5 \cdot 5) + (3 \cdot 3 \cdot 3)$$
$$153 = 1 + (1 \cdot 2) + (1 \cdot 2 \cdot 3) + (1 \cdot 2 \cdot 3 \cdot 4) + (1 \cdot 2 \cdot 3 \cdot 4 \cdot 5)$$
$$153 = (1 + 2 + 3 + \ldots + 15 + 16 + 17)$$

The last interpretation, i.e., that 153 is the sum of the numbers 1 to 17, seems particularly attractive because the letters of the Hebrew word *tov* (*good*, as in *mazel tov*) have numerical values that add up to 17. By this interpretation, 153 conveys that a plethora of good things is in store for us at the end of time.

22 *David* corresponds to 14.

23 *Nero Caesar* corresponds to 666.

Even more famous is the passage in the Apocalypse that deals with the "number of the beast": "Here is wisdom. Let him that hath understanding count the number of the beast: for it is the number of a man; and his number is Six hundred threescore and six." Because Nero, the Roman emperor, is a particularly apt candidate for this satanic symbol based on his name in Hebrew script, the almost general consensus is that this riddle was leveled squarely at him. There is, however, at least one other candidate. Diocletian, also a notorious persecutor of Christians, had the name and title DIoCLᴇs ᴀVɢVsᴛVs; contained within it is a set of Roman numerals:

$$(D + I + C + L + V + V + V) = (500 + 1 + 100 + 50 + 5 + 5 + 5) = 666.$$

Plato wanted his republic to be peopled by precisely 5040 inhabit-
ants, no more and no less. This is the product of the first seven numbers:
$(1 \cdot 2 \cdot 3 \cdot 4 \cdot 5 \cdot 6 \cdot 7) = 5040$. Why Plato chose as the ideal population figure
the product of the first seven numbers instead of the first six or eight or
ten is a moot question, especially because Plato must have been aware that
he was courting disaster with this specification. If one out of his 5040 citizens—a number with many divisors—died unexpectedly overnight, the
philosopher-king would be stranded with 5039 citizens—a catastrophe in
numerological terms, because 5039 is a prime number and has only two
divisors: one and itself.

It has often been asserted that the vacuity of number symbolism, which
has been apparent in the last two examples, is the result of giving fantasy
too loose a rein, of attributing more meaning to numbers than was ever
there to be found. The population figures of states are straightforward
indicators of their sizes. Whether they have many divisors or only two is
completely irrelevant to this function. For the Pythagoreans, however, the
"dumbing-down" of the concept of numbers to mundane quantification
was a sacrilege, a base and criminal act.

In this assessment, we may well decide to keep them company. For
Pythagoras, the infinite variety of the world and the intricate complexity
of cosmic processes can be understood only by delving down to their basic
building blocks, their axioms, their very constituent atoms—numbers.
Mere counting—simply supplying answers to "how many" questions—
will not make us privy to such understanding and cannot even begin to
tap the potential that lies dormant in numbers.

If truth be told—and the following pages will attempt to do just that—
the problem with number symbolism is not so much that it tends to flood
numbers with an embarrassing wealth of fantasy. On the contrary, its pat
numerological recipes often fall far short of Pythagoras' expectations and
bring a serious fantasy deficit to its encounters with the world.

Bach

Numbers and Music

When we listen to one of the great compositions by Johann Sebastian Bach—or Haydn, Mozart, Beethoven, Schubert or any other great composer—what strikes us is the depth of musical expression, the inspired ideas and, perhaps most of all, the consummate workmanship. The internal balance of these works is often so delicate that changing even a single note generates discord and leaving out a single phrase brings the whole edifice to the point of collapse.

24 Johann Sebastian Bach.

Rationalist analysis is not the proper tool for an exploration of the aesthetic perfection of a musical composition by a genius. Such an approach is no more effective than trying to explain the overall power of a work of art by using a wealth of technical vocabulary. Overall power and aesthetic genius are properties of the finished work; these we will be content to enjoy. As mathematicians, however, we can ask whether abstract numbers are capable of throwing at least a tentative ray of light on the building blocks that Bach used, consciously or unconsciously, to construct his deeply religious works.

The age in which Bach lived was undeniably very cognizant of the symbolic significance of numbers in the context of religious cult, and it soon becomes obvious that numerical relationships played a significant role in Bach's work. Bach's major works do not contain a single note that could be eliminated or that was added without due deliberation. The question is, can we demonstrate that he used the numerical relationships that exist between sequences of notes to weave statements of a predominantly religious nature into his compositions?

The primary source from which Bach drew not only the background knowledge required for a presentation of religious ideas in a numerological-symbolist disguise but also the motivation for such a presentation was undoubtedly Martin Luther's translation of the Bible. In his copy of the Bible, which included the commentary of the seventeenth-century theologian Abraham Calov (1612–1686), Bach underlined all passages featuring numbers that concern people or events.

In this interpretation, God is seen as "using compasses" when he makes his creation conform to the law of numbers and when he conveys the measurements for the Ark of the Covenant to Moses. This view also says that the divine master-builder placed numbers at Bach's disposal as building blocks for a musical tabernacle and that the points of Bach's own "compasses" can be detected in its fabric.

A second source for Bach was the philosophy of the enlightenment, notably the rationalistic thinking of Gottfried Wilhelm Leibniz (1646–1716). Among his many other accomplishments, Leibniz was an inspired mathematical genius of great originality who will be discussed in his own right in Chapter 5. His influence can be detected particularly in Bach's skillful permutation of notes, in which the composer used as his guide Leibniz' work *De arte combinatoria (On the Art of Combination)*, published in 1666 when Leibniz was barely 20 years old.

One permutation that is often found in Bach's music is that of A, Bb, C and B. It occurs in several forms, notably the *ascent* A–Bb–B–C, the *descent* C–B–Bb–A and the *cruciform* Bb–A–C–B. The cross has its own significance, but what makes these permutations so important is that, in German notation, B♭ is represented by *B*, and B by *H*. To Bach, therefore, this batch of notes was not A, B♭, B and C, but A, B, H and C, the letters of his own name: a musical signature that can be heard again and again.

One particularly famous instance is the final counterpoint in *Die Kunst der Fuge (The Art of the Fugue)*, in which B♭–A–C–B is the very last theme of the fugue. That this theme is rounded off with C♯–D, the next notes on a rising scale after those of Bach's "signature", hints at the theme of *exaltation*. Looking in Bach's Calov Bible, we duly find that he has underlined the following phrase: "Humble yourselves therefore under the mighty hand of God, that he may exalt you in due time."

Another example is the first bar of the A minor prelude from Part II of *Das Wohltemperierte Klavier (The Well-Tempered Clavier)*:

The treble starts with a sequence expressive of pain and incorporates the four "signature" notes in a descending sequence of semitones (C–B–B♭–A), to which the bass adds six notes in a sequence of descending semitones A–G♯–G–F♯–F–E. There are ten notes in all, corresponding to the number of the divine commandments.

Worthy of a rather more detailed discussion are the numerical relationships encompassed in the theme of the last fugue from Part I of *The Well-Tempered Clavier*, the one in B minor.

This theme is audibly related to the "Kyrie" of the Mass in B minor. It starts with an arpeggiated B minor triad. The next 12 notes sigh their way through a series of six stepped minor seconds until the theme ends with an arpeggiated F♯ triad and the return to the dominant of F♯. We find a closely related theme being played by the viola da gamba in the prelude before "*Es ist vollbracht*" ("It is finished"), in *The Passion According to St John*. Again—and significantly—the prelude is in B minor.

Once again, there is a numerological connection. Can it be an accident that this prelude is 19 bars long and that the fugue in B minor from *The Well-Tempered Clavier* is exactly 4 · 19 = 76 bars long?

Also remarkable is the fact that this fugue consists of 14 repetitions of the theme. The addends of 14 (2 + 1 + 3 + 8) can be interpreted as B + A + C + H, if every letter is assigned the number that it occupies in the alphabet. His signature again, running right through the final fugue—a fitting seal for that monumental opus, *The Well-Tempered Clavier.*

However, we are by no means finished with this remarkable piece of music. Another striking fact about this fugue is that its theme uses all 12 notes of the chromatic scale.

Twelve: if we return to numerology, we find that this number is rich in symbolism. There are *twelve* notes in the chromatic scale, *twelve* signs of the zodiac, *twelve* months in a year. *Twelve*—the product of *three* and *four*, each of them steeped in its own mystery—*three* persons of the trinity, *four* points of the compass, *three* spatial dimensions, the *four* elements of antiquity (earth, air, fire and water) and so on.

To return to the theme of the B minor fugue, it is true that the twelve tones do not occur only once each, as they would in a Schoenbergian twelve-tone row. It is also true that the frequency with which they occur is different for different notes. The note occurring most frequently (five times) is F♯, the dominant to the tonic of B. Of the other notes, five are below F♯ and eleven above it.

Five and *eleven*: in numerology, these numbers usually appear in a tragic context: the Five Wounds of Christ come to mind, and the fact that *eleven*, as the *transgression* of *ten* (cf. the Ten Commandments), symbolizes *sinful* transgression.

One last comment on *twelve*: in Bach's number symbolism, switching numbers around in two-digit figures often amounts to a reversal of what they originally implied as symbols. If *12* stands for the perfection of Creation, its reversal, *21*, symbolizes the yearning for redemption, and there are 21 notes in the theme of the fugue.

Enough! Fascinating as it is to meditate on numbers and their heavy cargo of symbolism, such "science" must always remain inconclusive, doubt-ridden and superficial. It is also restricted to a single, if particularly scintillating, facet of numbers. The true nexus between numbers and music reaches to depths of an entirely different order.

Let us consider what happens when we hear a single note. Our ears perceive a periodic sequence of vibrations. In fact, the note generated by a tun-

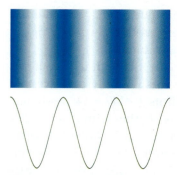

25 Sine wave: as a note is given, the air is compressed in a periodic pattern temporally and spatially.

ing fork or some "hi-tech" device, such as an electronic sine-tone generator, enters our organ of hearing as an elementary sine wave. Its frequency (the frequency with which maximum and minimum air pressure alternate per second) is measured in hertz.

Before the conductor turns up, the oboist of an orchestra (it has to be the oboist because the oboe is the only orchestral instrument incapable of being tuned on the spot) produces a 440-hertz note, the chamber tone A1. The other members of the orchestra then use this note to tune their instruments. Needless to say, the oboe with its nasal charm gives a much more pleasant sound than the curiously empty, "naked" sound produced by a tuning fork or a sinus generator.

Why is this so? Simply because the sound of an instrument, even if the musician only plays one single note, does not arrive at our ears stripped down to a basic sine wave. Rather, it comes in the merry company of what are known as its "overtones": those notes whose frequencies are exact multiples of the frequency of the fundamental.

It is a remarkable fact that overtones were unknown in antiquity; in fact, they were discovered as recently as 1636 by the French mathematician, scientist and theologian Marin Mersenne (1588–1648) and not studied in detail until 1702, by his compatriot, the physicist Joseph Sauveur (1653–1716). It was 1878 before the physical properties of overtones were exhaustively discussed by John Strutt, 3rd Baron Rayleigh (1842–1919) in *The Theory of Sound*, a work that retains its position as a classic of acoustics to this day.

The degree to which these overtones enrich their fundamentals is responsible for the specific timbre of a musical instrument and, indeed, of the human voice, where it is also the decisive factor for keeping the vowel sounds discrete. It was one of the most remarkable achievements of mathematics in the nineteenth century when Joseph Fourier (1768–1830) discovered that practically *any* wave form—and therefore the

26 Joseph Fourier.

timbre of any instrument—could be realized by stacking overtones on their fundamental. Electronic synthesizers have been spectacularly successful in incorporating Fourier's discovery.

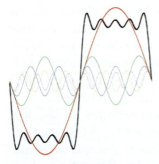

27 A note is the combination of the fundamental frequency and the frequencies of its overtones.

Let us now consider in greater detail the theory that underpins this particular nexus between *number* and *sound*.

If a pianist strikes a D with its frequency of 147 hertz, our ears not only perceive this fundamental with its 1 · 147 = 147 hertz but also the D an octave higher with 2 · 147 = 294 hertz, the A above it with 3 · 147 = 441 hertz, the ensuing D with 4 · 147 = 588 hertz, the ensuing F♯ with 5 · 147 = 735 hertz, the ensuing A with 6 · 147 = 882 hertz, the ensuing C with 7 · 147 = 1029 hertz, the ensuing D above it with 8 · 147 = 1176 hertz and so on. It is not the frequencies of each sound that are important for us—these are defined by the chamber tone which is pegged on mere convention—but the *factors* that emerge from them, i.e., the numbers in sequence from one. What these impart to us are the *ratios* of the frequencies of all the sine waves that will form, as the fundamental (in our example, the note D on the piano) is sounded.

28 The overtones of D on a piano keyboard.

This all sounds very precise; however, note that almost everything just said is only approximately correct. It cannot be otherwise, as we shall now see as we examine the causes of this imprecision in greater detail before finally returning to our fugue theme from *The Well-Tempered Clavier*.

First of all, let us get a few things straight.

The ratio of the frequencies of two tones—a fundamental and a second tone which represents a step either up or down in pitch and is sounded either together with the fundamental or immediately after it—is called a *musical interval*. This ratio of frequencies can be expressed numerically.

The most elementary interval is the *prime*, in which a fundamental relates to itself. The numerical ratio here is 1:1. The next interval, in an ascending order of complexity, is the *octave*, in which the fundamental relates to the note with double its frequency; the corresponding numerical ratio for this interval is 1:2. This note is the first in the series of *overtones*.

What has been said of the octave is true without any further qualifications. If we treat the higher tone of the octave as a new fundamental, its overtone series would consist exclusively of overtones of the original

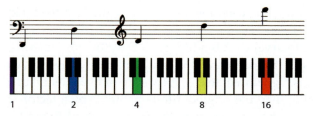

29 Octaves with the double, quadruple, octuple, ... frequencies of the root become audible.

fundamental—apart, of course, from the prime of the original fundamental itself. The octave is the second most consonant interval after the prime. This is because our ear experiences all of the sounds generated by these two tones as belonging together in an ideal way.

The affinity of the two tones of an octave is so close that they sound the same to us. In other words, two tones are perceived as equivalent if the frequency of the higher one has been obtained through the doubling of the frequency of the lower one. We can summarize as follows:

> A musical interval shall be considered to have remained unchanged, if it is multiplied by two (i.e., if we play the second tone an octave higher or the first tone an octave lower).

> A musical interval shall equally be considered to have remained unchanged if it is divided by two (i.e., if we play the second tone an octave lower or the first tone an octave higher).

From this agreement, it follows that the numerical ratios of all musical intervals are limited in such a way that they amount to at least one but are smaller than two. Musically speaking, it follows that the second tone of an interval (taking its octave as required) sounds at least as high as the first tone but lower than its octave.

Let us consider the overtone series of D as an example. The overtone that follows after the octave is A—three times the frequency of D. If we play this A an octave lower, the resulting interval D–A corresponds to the numerical ratio of 3:2. This interval is designated a *fifth*; every second overtone of A coincides with every third overtone of D.

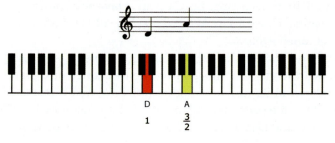

Pythagoras and his school hatched the idea of using the fifth to obtain further intervals.

The *second*—9:8. The interval D–A is a fifth with D as the fundamental; A–E is a fifth with A as the fundamental. Put mathematically, this tells us that the frequency of E is higher than that of A by a factor of 3/2. Similarly, we know that A is higher than the original fundamental D by a factor of 3/2. So what is the factor when we compare the frequencies of E and D? It is simply a matter of multiplication. The frequency ratio of E and D is (3:2) · (3:2) = 9:4. Our next step is to transpose E down an octave. How does this affect the ratio? Well, you will remember that the frequency of the note one octave below the fundamental is one half that of the fundamental—ratio 1:2. Therefore, the ratio of the transposed E and the original D is arrived at by the equation (9:4) · (1:2) = (9:8). An interval with such a ratio is known as a *second*.

The *sixth*—27:16. The interval E–B, like D–A and A–E, is a fifth, this time related to the fundamental E. The frequency of B is higher than that of E by a factor of 3/2. We can therefore arrive at the frequency ratio of B and D by multiplication: (E–B) · (D–E) = (3:2) · (9:8) = 27:16. This interval is known as a *sixth*.

This yields the following table of intervals and numerical ratios:

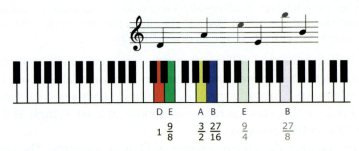

The *fourth*—4:3. Pythagoras goes on to look for the fundamental G with which D forms a fifth. Because the frequency ratio of G–D is that of a fifth, 3:2, the reciprocal frequency ratio D–G must correspond to its inverted form of 2:3. The G that is an octave above has a frequency double that of its fundamental. By transposing this G up an octave and sounding it together with the original D, we arrive at the ratio (2:3) · (2:1) = 4:3. An interval with this ratio is called a *fourth*.

The *seventh*—16:9. As before, it is now possible to look for the fundamental C with which G forms a fifth. The frequency of C stands in ratio 2:3 to G, which in its turn was higher (after we had transposed it up by an octave) than the original fundamental D by a factor of 4:3. The ratio of frequency of C and D is therefore (2:3) · (4:3) = 8:9. If we then transpose this

C an octave higher, it stands in ratio $(2{:}1) \cdot (8{:}9) = 16{:}9$ to the fundamental D. This D–C with its numerical ratio of 16:9 is an example of the musical interval known as a *seventh*.

The *third*—32:27. Finally let us consider the fundamental F with which C forms a fifth; the frequency of F is lower than that of C by a factor of 2:3. The frequency ratio of F and D is therefore $(2{:}3) \cdot (16{:}9) = 32{:}27$, corresponding to our penultimate musical interval, the *third*.[14]

We may now draw up the following list of intervals and numerical ratios:

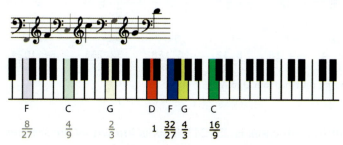

In this way, the Pythagoreans obtained, within an octave scale, the seven notes D, E, F, G, A, B, C, which now correspond to the white keys on a piano.

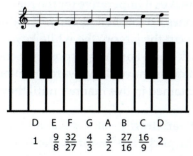

The *semitone step* or *the minor second*—256:243. The smallest difference in this scale is between the notes E and F, and between B and C. The interval E–F is computed as follows: D–E is a second: so we multiply the frequency of E by 8:9 to find the frequency of D. The interval D–F is a third so multiplying the frequency of D by 32:27 gives us the frequency of F. Therefore, the interval E–F has the following ratio:

$$(8{:}9) \cdot (32{:}27) = 256{:}243.$$

Following a similar process, we can compute the ratio for the interval B–C (the inversion B–D of the sixth D–B multiplied by the seventh D–C):

$$(16{:}27) \cdot (16{:}9) = 256{:}243.$$

D	A	E		B		F♯		C♯		G♯	
1	$\frac{3}{2}$	$\frac{9}{4}$	$\frac{9}{8}$	$\frac{27}{8}$	$\frac{27}{16}$	$\frac{81}{16}$	$\frac{81}{64}$	$\frac{243}{32}$	$\frac{243}{128}$	$\frac{729}{64}$	$\frac{729}{512}$

30 Ascending fifths from D to G♯.

D	G		C		F		B♭		E♭		A♭	
1	$\frac{2}{3}$	$\frac{4}{3}$	$\frac{4}{9}$	$\frac{16}{9}$	$\frac{8}{27}$	$\frac{32}{27}$	$\frac{16}{81}$	$\frac{128}{81}$	$\frac{32}{243}$	$\frac{256}{243}$	$\frac{64}{729}$	$\frac{1024}{729}$

31 Descending fourths from D to A♭.

The result in both cases is 256:243. This is known as the *semitone step* or *minor second*, which is such an extraordinarily dissonant interval that only every 256th overtone of the lower tones (E, B) coincides with every 243rd overtone of the higher ones (F, C).

The semitone step makes it imperative for the remaining semitones to be added to the whole tones established already. The diatonic scale becomes a chromatic one or, to cut the jargon, the white keys of the piano are supplemented by the black ones. For this purpose, we continue to stack fifths one on top of the other or move downward in intervals of fourths and transpose the tones obtained in this manner until they fall within the compass of one octave.

If you stop short—arbitrarily—at G♯/A♭ and arrange the tones in sequence according to pitch, the result is the intervals of Pythagorean tuning.

In this scale, the tones A♭ and G♯ are extremely close to each other. The interval G♯–A♭ is the numerical ratio obtained when the interval A♭–D (the reciprocal value of the interval D–A♭) is multiplied by the interval D–G:

$$(729:1{,}024) \cdot (729:512) = 531{,}441:524{,}288 = 1.0136.$$

In other words, this interval differs from the prime only by 0.01369 (≈ 1.4%)—very much less than the 5.3% difference found in the semitone step (256:243 ≈ 1.053497). This difference (1.4% between G♯ and A♭), which is not necessarily noticeable even for a musician, has been called the *Pythagorean comma*.

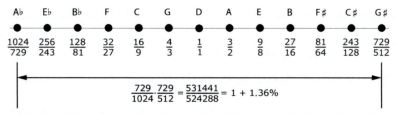

32 Twelve fifths stacked on top of each other (from A♭ to G♯) add up to almost exactly seven octaves.

In view of its comparative insignificance, the creators of keyboard instruments decided not to enrich the scale by stacking more fifths on top of each other—or, to put it in visual terms, not to add further black keys to the existing keyboard. This is one of the instances of *imprecision* we mentioned earlier. The pianist does not think twice about committing the sin of *enharmonic change* and quite simply equates G♯ with A♭, D♯ with E♭, A♯ with B♭, etc., even though, strictly speaking, this is not permissible.[15] No such license is permitted to the violinist because the strings of this instrument allow far greater precision.

For the modern musician, a more disturbing anomaly in the Pythagorean scale is the dissonant nature of the major third, D–F♯. Pythagorean analysis yields for this interval a numerical ratio of 81:64: dissonance indeed, as by this reckoning only every 81st overtone of D coincides with every 64th overtone of F♯. Clearly, this is not the case. An F♯ occurs in the overtone series of D as the fourth tone with the fivefold frequency of the fundamental. If you transpose this overtone down two octaves, we get the consonant major third D–F♯ with the numerical ratio of 5:4. Every fourth overtone of F♯ coincides with every fifth overtone of D.

We may compute the difference between the consonant major third and the (slightly augmented) dissonant Pythagorean major third by analogy

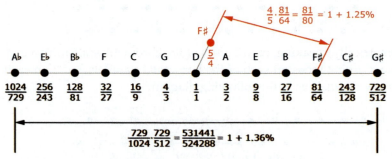

33 The Pythagorean third differs from the just third by 80:81.

with the interval A♭–G♯ discussed earlier. The reciprocal of 5:4 is multi-plied by 81:64:

$$(4:5) \cdot (81:64) = 324:320 = 81:80 = 1.0125.$$

This difference of 1.25 % compared to the prime is called the *syntonic comma*.

Can the syntonic comma itself be avoided? Yes, but only by abandoning altogether the Pythagorean idea of obtaining tones exclusively by means of fifths and fourths. Pythagorean tuning, one must remember, is one-dimensional in that it relies exclusively on fifths and on fourths (which are characterized by the prime number three). This is why the numerators and denominators of these intervals are divisible only by two and three. Mod-ern European music has defined 5:4 as the ratio for the major third — reject-ing the Pythagorean 81:64 — and thus obtains tones in *harmonic* or *just into-nation*, which is two-dimensional. All tones are obtained either through fifths (characterized by the prime number three) or through major thirds (characterized by the prime number five).

To illustrate this, we plot the fifths, starting with A♭ and ending with G♯, along a line in such a way that our exemplary D appears in the mid-dle. This gives us A♭–E♭–B♭–F–C–G–D–A–E–B–F♯–C♯–G♯. Next, we plot the two tones B and F♯ on a line parallel to the first one but one level above it, choosing the positions of these two tones in such a manner that they form an equilateral triangle with D below. We then complete the line by adding the fifths on either side of F♯ which gives us C–G–D–A–E–B–F♯–C♯–G♯–D♯–A♯–E♯. So far, so good. Our next step is to plot B♭ and F on a line parallel to the first one but one level below it, choosing the positions of these two tones to form an equilateral triangle with D above. Once more, we complete the line by adding the fifths on either side of B♭ and F, pro-ducing the sequence C♭–G♭–D♭–A♭–E♭–B♭–F–C–G–D–A–E.

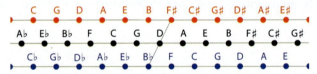

Geometrically speaking, the 12 tones of the chromatic scale form a par-allelogram with 4 tones on each of the horizontal parallels and 3 on each of the skew parallels as in the following diagram:

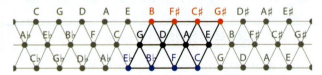

The numerical ratios of the consonant intervals, which trace a hexagon on the two-dimensional level, are ranked as follows:

The *prime* and the *octave* D–D.

The *fifth* D–A and the *fourth* D–G.

The *major third* D–F♯ and the *minor sixth* D–B♭.

The *minor third* D–F and the *major sixth* D–B. In these intervals the tones F and B have been defined in such a way that F–A and G–B form *major thirds*.

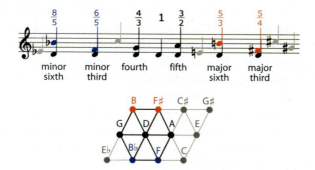

Just intonation imparts a harmonic character to the major triad D–F♯–A and to the minor triad D–F–A, which are, geometrically speaking, triangles with mirror-symmetry along the axis of the fifths. We can visualize the simple *cadence* of tonic D, subdominant G, dominant A and tonic D as a series of shifts of the "major triad" within the parallelogram.

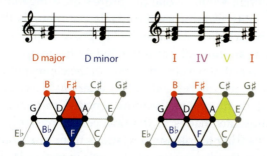

The notion that the fifth and the major third stake out the directions of two-dimensional music is capable of further theoretical refinement. Strictly speaking, the sixth overtone of D is not, as was said earlier, the C of the minor seventh (transposed down two octaves), it is lower than that note. It is, in fact, the tone with the sevenfold frequency of the fundamental, and transposed by those two octaves, it forms with D the interval of the natural seventh[16] with a numerical ratio of 7:4. If we use the 12 tones of the two-dimensional parallelogram in pure intonation as a basis on which to build natural sevenths, we get a three-dimensional space. This is filled with tones for which the prime numbers three (for fifths), five (for major thirds) and seven (for natural sevenths) are structural principles. The natural seventh is indeed used in some styles of modern music, such as jazz. In classical European music, it was considered an *ecmelic interval*—literally an interval absent from traditional melodies. This looks like an arbitrary decision in favor of reducing the dimensions of listening to two.

Integrating the natural seventh and the ecmelic tones generated by it into our tonal system would by no means establish "perfect" conditions for listening. As we move up in the frequency range, the next ecmelic interval would be found lurking in the shape of the overtone with the 11-fold frequency of the fundamental, which would throw open the door to a fourth dimension of music. Similarly, the overtone with the 13-fold frequency would call for a fifth dimension. It is obvious that every overtone corresponding to an odd prime number multiple of the frequency of the root would call into being a further dimension of the tonal system. Because there are infinitely many prime numbers, the tonal system is, strictly speaking, *infinite-dimensional*.

However, limiting the tonal system to two dimensions actually involves no great loss. Metaphorically speaking, it is only in the "ears of God" that music is infinite-dimensional. The human sense of hearing is restricted to finite sounds. Therefore, we have no choice but to accept a tonal system with a finite number of dimensions.[17] To compensate us for whatever loss that may entail, we have one of nature's most ingenious gifts at our disposal—the ability to improve intuitively on intervals that are ever so slightly "off". But what do we mean by "improve intuitively"?

When we hear an interval, we immediately try to assign it to its place among the intervals that we already have. In the context of European classical music, these are the *just intonation* intervals described earlier. The more consonant an interval is, the more sensitive our ears prove to be by picking up minuscule shades of dissonances. Conversely, the readiness of our sense of hearing to tolerate deviations from pure intonation grows proportionately as the dissonance of intervals increases.[18]

Even if we submit to the limitation of two dimensions when listening to music, we would, in order to assess pure intonation intervals with the "ears of God", still need an *infinite* number of tones to comprehend two-dimensional music. Equipped as we would then be with an "infinitely precise" sense of hearing, we must still be mindful of the Pythagorean comma. We succeeded in circumventing the syntonic comma by opening up a second dimension. It is, of course, still there (in addition to the F♯, which forms

the 5:4 major third with D, there is the unrepentant F♯ generated exclusively by fifths, which forms with D the 81:64 Pythagorean major third). As
the second dimension evolves, further "commas" are added. Three major
thirds stacked on top of each other can be summed as follows:

$$(5{:}4) \cdot (5{:}4) \cdot (5{:}4) = 125{:}64.$$

This is almost, but not quite, the 128:64 (= 2:1) of a full octave. We may
quantify the difference as 128:125 = 1.024. This deviation of 2.4% is known
as the *minor diesis*.

Similarly, four *minor* thirds stacked on top of each other add up to
slightly more than a full octave:

$$(6{:}5) \cdot (6{:}5) \cdot (6{:}5) \cdot (6{:}5) = 1{,}296{:}625.$$

The difference between this and the full octave's 1,250:625 may be expressed
as 1,296:1,250 = 1.0368.

This deviation of almost 3.7% is called the *major diesis*. As the Pythagorean comma disposes of enharmonic change, so the minor and the major
diesis join forces to prevent us from calling these tones (i.e., those generated by the addition of three major and four minor thirds) the pure note
D. *Just intonation* generates a lattice of an infinite number of notes based on
the note D unfolding on an unbounded plane that is nonetheless encapsulated within a single octave. This was first formalized mathematically by
the eminent Swiss mathematician, Leonhard Euler (1707–1783).[19]

34 Leonhard
Euler.

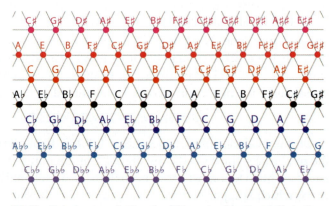

35 The Eulerian tone lattice: the fifths extend horizontally to the
right, the fourths horizontally to the left, the major thirds diagonally upward to the right, the minor sixths diagonally downward
to the left, the major sixths diagonally upward to the left and the
minor thirds diagonally downward to the right.

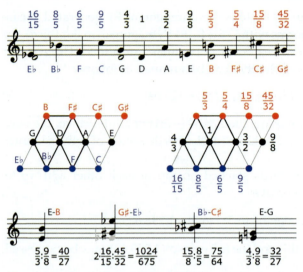

36 The fifth E–B, the fifth G♯–E♭ (notated as a diminished sixth), the minor third B♭–C♯ (notated as an augmented second), and the minor third E–G sound out of tune when just temperament is based on the root D.

Neither the Pythagorean comma nor the minor diesis nor the major diesis can be called good news for the piano. True, we may find it quite handy to deal with 12 notes per octave rather than the infinite number found in Euler's lattice. There is much to be said, after all, for concentrating our gaze on a tightly circumscribed parallelogram of 12 notes rather than letting it wander over Euler's unbounded plane. What is not tolerable, however, is that if the instrument is tuned on the just intonation model with a base (for example, D), all intervals that are not directly related to this base are more or less out of tune. The fifth E–B, for instance, is based on the calculation (8:9) · (5:3) = 40:27. As (3:2) · (27:40) = 81:80; this so-called *wolf fifth* differs from the pure fifth E–B by the 1.25% of the syntonic comma.

Similarly, the major third G♯–C is calculated as follows:

$$(32:45) \cdot (9:5) = 288:225 = 32:25.$$

On the evidence of (32:25) · (4:5) = 128:125, this interval deviates from the pure major third G♯–B♯ by the 2.4% of the minor diesis and is experienced as much less satisfactory even than the dissonant Pythagorean major third, which is separated from the pure major third only by the syntonic comma. As we have said, the syntonic comma amounts to only 1.25%, or just over half the value of the minor diesis.

The piano has survived as a musical instrument courtesy of the fact that we are a long way from listening to it with the "ears of God".

As long ago as 1585, Simon Stevin (1548–1620), a Dutch universal genius in the true Renaissance mold, whose many credits include the introduction of decimals into Western mathematics, came up with an ingenious formula for tuning keyboard instruments. Stevin's formula relies on the ability of our sense of hearing to adjust intervals as we process them. Putting his decimals through their paces, Stevin defined the difference in frequency of the minor second as the twelfth root of two. This is known

37 Simon Stevin.

as an *infinite decimal*, 1.059463094…, which has the remarkable property of yielding the product 2 when multiplied by itself 12 times.[20] This number or, strictly speaking, a finite decimal number that represents a sufficiently close approximation to it, was posited by Stevin as the difference in frequency between any two successive notes on the chromatic 12-note scale of the piano. The theory is clear: each octave contains 12 minor seconds, so 12 minor seconds stacked on top of one another result in one octave *precisely*. In the following table, Stevin's *equal temperament* is compared to *just intonation*.

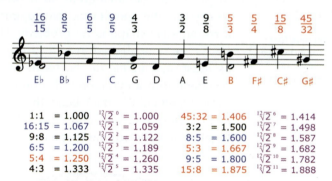

1:1 = 1.000	$\sqrt[12]{2}\,^0$ = 1.000	45:32 = 1.406	$\sqrt[12]{2}\,^6$ = 1.414
16:15 = 1.067	$\sqrt[12]{2}\,^1$ = 1.059	3:2 = 1.500	$\sqrt[12]{2}\,^7$ = 1.498
9:8 = 1.125	$\sqrt[12]{2}\,^2$ = 1.122	8:5 = 1.600	$\sqrt[12]{2}\,^8$ = 1.587
6:5 = 1.200	$\sqrt[12]{2}\,^3$ = 1.189	5:3 = 1.667	$\sqrt[12]{2}\,^9$ = 1.682
5:4 = 1.250	$\sqrt[12]{2}\,^4$ = 1.260	9:5 = 1.800	$\sqrt[12]{2}\,^{10}$ = 1.782
4:3 = 1.333	$\sqrt[12]{2}\,^5$ = 1.335	15:8 = 1.875	$\sqrt[12]{2}\,^{11}$ = 1.888

The table shows that equal-tempered intonation defines intervals so precisely that a musically trained ear processes them as perfectly tuned intervals. It so happens that this type of intonation is particularly good at rendering the consonant intervals of the fourth and the fifth, to which our ear is especially sensitive—a singular stroke of good luck.

Equal-tempered intonation gets rid of not only the Pythagorean comma but also both the minor and major dieses. The infinite number of tones, which Euler mapped on the lattice expanding in an unbounded plane, is reduced to the 12-note scale of the piano. To visualize this reduction, we can use a *torus*—an attractive geometrical model in which the plane assumes

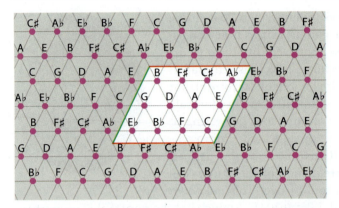

38 In well-tempered intonation, the infinite number of notes of the Eulerian lattice is reduced to the 12 notes of the chromatic scale.

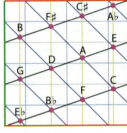

39 The parallelogram of the 12-tone scale has been distorted here to a rectangle. Gluing the two green and the two red sides together results in a torus.

the shape, broadly speaking, of a doughnut. The 12 tones of the piano can be visualized as being evenly distributed on its surface.[21]

To make the torus take shape in our mind, we proceed as follows: we take the Eulerian lattice and cut out a parallelogram containing the 12 tones of the chromatic scale. We then proceed to print this parallelogram on a rubbery foil so expandable and elastic that tugging it into a rectangular shape is no problem. The major thirds are now aligned in four groups of three as illustrated in Figure 39. Next, we roll this rectangle into a cylinder and glue its adjacent sides together. We then bend this cylinder to form a closed surface and again glue its ends together to produce the shape of a ring dough-

40 The rotoid connects the 12 notes of the chromatic scale along the circle of fifths Ab–Eb–Bb–F–C–G–D–A–E–B–F♯–C♯ and circles the torus three times in doing so. The 4 circles inscribed on the torus, D–F♯–Bb, A–C♯–F, and B–Eb–G, each connect 3 notes according to the major-third relationship.

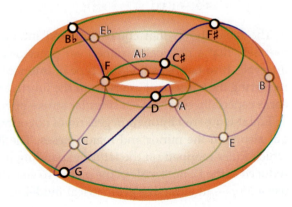

nut. We can now see a curve that meanders around our doughnut three times and connects the 12 tones in the sequence of the circle of fifths, D–A–E–B–F♯–C♯–A♭–E♭–B♭–F–C–G (–D). This curve is known as a *rotoid*.[22] From a mathematical point of view, the pianist plays the notes in a set of points on a path wrapped around the torus.

Bach was utterly taken by the idea of well-tempered intonation. He did not yet use Simon Stevin's equal-tempered variety described earlier. Stevin's system can be improved on only marginally with electronic tuning devices. We cannot ask perfection of a system based on the twelfth root of two, a number whose own value can only ever be expressed approximately. The system preferred by Bach for his famous *The Well-Tempered Clavier* was that of Johann Philipp Kirnberger (1721–1783), a viable realization of well-tempered intonation. One of his aims in so doing was to demonstrate how easily keyboard instruments tuned in this way can accommodate changes of key. This brings us back at last to the theme of the last fugue in *The Well-Tempered Clavier, Part I*.

It is our ability to process intervals as pure even when they are not (as is consistently the case in well-tempered scales) that allows us to sidestep the limitations of keyboard instruments and move toward an intuitive divination of what music sounds like in the "ears of God". In the theme of the last fugue of *The Well-Tempered Clavier, Part I*, if we follow the succession of intervals through, the final F♯ is by no means identical with the initial one. In fact the two, far from having the ratio 1:1 = 1, are a long way apart on the Eulerian lattice.[23] This succession of intervals corresponds to repeated orbits of the 12-tone torus, and the resulting interval computes as

15,625:16,384 = 0.95367.

When we have accommodated the tones that we hear to just intonation, we perceive the theme's final F♯ to be more than 4.6% lower than the first one.[24]

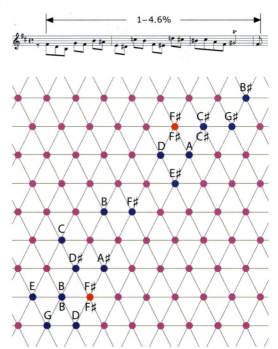

The pianist has no option but to strike the same key again and again, yet the F♯ at the end of a melody that is haunted by unfulfilled longing force-fully presents itself to our ears as shrouded in darkness. Did Bach bring this about intentionally, or is it the work of his intuitive genius? We shall never know. The area on the musical map occupied by this fugue should perhaps bear the legend: *Here dwell secrets.*

Hofmannsthal

Numbers and Time

When we consider the mysteries that surround us, which of them is more baffling than time? We study a fading photograph of one of our forebears. Where has that smile gone? Where's that glance at the camera, so shy and yet so full of expectation? What has become of that half-articulated yearning for acceptance? We look into the eyes of a little child as she stretches out her arms to us; we gaze at her face transfigured with the joy of recognition. What does the future have in store for her? How will she cope

41 Hugo von Hofmannsthal.

with unrequited love or with obstacles that block her ambitions? What decisions will shape her life after we ourselves are gone? Or again, at a moment of supreme happiness, we may wish to emulate Faust and cry out to the fleeting moment, "Linger yet, you are so beautiful!" We need not fear Mephistopheles' fetters. We can say what we like—there is not the slightest chance of our wish being granted. Nothing that we can do can ever stop time, even for the blink of an eye.

The future is forever out of our reach, the present slips away from us faster than thought, the past is beyond recall. Death will take away our power to influence the future, but our responsibilities remain. We are haunted by not knowing what tomorrow will bring, haunted by the constant fear that we are even now missing out on opportunities that tomorrow may prove to have been all-decisive. We are filled with regret at the way happy memories fade swiftly into distant echoes, never to be recalled. Time's yoke—how keenly we feel its weight on our protesting shoulders!

The poet Hugo von Hofmannsthal (1874–1929) was barely twenty when he addressed the mystery of transience in his *Terzinen: Über Vergänglichkeit*.

> Stanzas in "Terza Rima: On Transience"
> (Translated by Michael Hamburger)
>
> Still on my cheek I feel their warm breath fall:
> How can it be that these near days are spent,
> Past, wholly past, and gone beyond recall?
>
> This is a thing that mocks the deepest mind
> And far too terrifying for lament:
> That all flows by us, leaving us behind.
>
> And that unhindered my own self could flow
> Out of a little child whom now I find
> Remote as a dumb dog, and scarcely know.
>
> Then: that in lives a century old I share
> And kinsmen laid in coffins long ago
> Are yet as close to me as my own hair,
>
> Are no less one with me than my own hair.

Hofmannsthal seems to have worried at this theme all through his life. Fifteen years later, just five years before the outbreak of World War I, he wrote the libretto for the comic opera *Der Rosenkavalier*. In Act 1 of this work, we find the Marschallin contemplating her face in her hand-mirror and uttering the following moving words:[25]

> But how can it be
> That I was little Tess
> And soon, so very soon

I shall be old?

. . . .

How can that be?
How can the good Lord above allow it?
I know I've always been the same.
And if it has to be this way,
Why do I have to know
With my mind wide open?
Why do I have to see
So clearly?
Why does He not keep me in the dark?
Such a mystery,
Such a very great mystery.

Individuals, finite and transient as they are, find many ways to revolt against this universal human condition. One of the most remarkable of these is simply to turn your back on the undeniable fact that we are all defenseless against the march of time. Taken to its extreme, this approach leads its practitioners to live in complete denial of the inescapable fact that there was a time when they were not and there will be a time when they will be no more.

The Greek philosopher Parmenides of Elea (born c. 515 BC) was the first thinker to try to construct a logical basis for this rejection of transience. His answer was to declare categorically that the notion of time was an illusion.

Many of the earliest Greek philosophers are little more than names to us—associated at best with a few disparate fragments. Parmenides is different. Elea was a Greek colony in southern Italy, and Parmenides' life straddles the sixth and fifth centuries BC. It is probable that he

42 Parmenides of Elea.

was for a time under Pythagorean influence. Substantial parts of a didactic poem have come down to us. We see Parmenides grappling with such riddles as why day and night follow one another and why the sun is visible for different periods at different times of the year. We see him discussing how the various celestial bodies were formed from fire so far from the earth and how the elements maintain equilibrium on the earth. Throughout, we see him drawing on insistent logic to prove that the notion that anything substantial can either be created from nothing or decay into nothing is untenable and that any discourse based around the ideas of coming into being and ceasing to be is intrinsically self-contradictory. For Parmenides, movement is mere appearance, change illusory and the

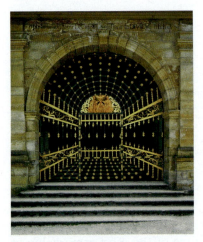

43 A *trompe l'oeil* perspective creates an impression that is found to be an optical illusion.

actions of coming into existence and ceasing to exist over time quite simply impossible.

To substantiate this theory, Zeno of Elea (c. 495–430 BC), a pupil of Parmenides, created his famous *arrow paradox*. Imagine an arrow in flight. At any one moment it must occupy a discrete position in space; at any one moment, it must be standing still at a clearly defined locus. Yet if, at any one moment, the arrow is standing still, how does it ever manage to change its position?[26]

One answer is to appeal to the testimony of our eyes that the arrow in fact moves rapidly in flight. Zeno's response may have been something like this:

> Another instance of how our senses deceive us! Have we not had a similar experience when our eyes tell us that a straight rod partially immersed in water is crooked? The only evidence we can rely on as unassailable is evidence generated by logical argument.

Piling rhetoric upon rhetoric, he may have continued:

> Which do you think is more deserving of our trust? Sensory perceptions, notorious as they are for leading us astray with all kinds of deceptions and hallucinations, or the irrefutable conclusions of transparent, guileless logic?

Aristotle (384–322 BC) was the first to point out the flaw in Zeno's seemingly irrefutable conclusions. Zeno's paradox required not only that the finite time span of the arrow's apparent motion be divided into an infinite number of discrete moments but also that the finite distance ostensibly traveled by the arrow be divided in its turn into an infinite number of discrete loci. Without these two premises, Zeno could not claim that at any one moment the arrow must be standing still at a clearly defined locus. But where did he get these premises? There is no logical reason for us to treat infinity in such a

44 Aristotle.

manner. In fact, as Aristotle quite rightly pointed out, all human experience flatly contradicts any such assumption—nowhere in real life do we encounter infinity.

This is as true for us as it was for Aristotle. The world is finite even under the most powerful microscope, and the indeterminacy principle within quantum mechanics blocks any project to subdivide a distance AB into infinitely many loci. At the other end of the scale, the idea of outward bound infinity founders at the cosmic horizon beyond which no signal can penetrate. The far side of this horizon can have no relevance for us. Even computers cannot handle infinity. They are incapable even of infinite regression, a task they will terminate after a finite number of operations, and they make use of a finite number of both symbols and pixels.

So much for Zeno's attempt to give a logical basis to Parmenides' assertion that time is nothing more than a sensory delusion. We must admit that Parmenides' denial that time has any real existence leads us ultimately to a dead end. Where does that leave us? Still asking the same question, "What is time?"

It appears we are dealing with ideas that seem to be mutually exclusive: on the one hand, development and decay; on the other, permanence and immutability. We will see that, far from being mutually exclusive, these ideas complement each other—in fact, neither pair is conceivable without the other. If we want to arrive at a mental construct of time, we must first strike a balance between transience and perpetuity. What enables us to do this we will call *rhythm*—a concept that ancient thinkers described metaphorically as the *Eternal Recurrence of the Same*. So, what do we mean by *rhythm*? We must answer this question before we continue our investigation into the nature of time.

From the beginning of our history, human beings have experienced rhythm in the cyclical movements of the sun, the moon and the stars. At the very root of our sense of time lies the seemingly eternal recurrence of these three clusters of celestial phenomena.

First, there is the sun. It rises every morning, reaches its apex at midday and sets every evening. This dance is complemented by the apparent movements of the so-called fixed stars. We now know that all these effects owe their existence to the earth's rotation on its own axis. We call the period of one full rotation a *sidereal day*—a unit of time more or less identical to the period between two successive apexes of the sun at midday.[27]

When we use the terms *day* and *night* in opposition, we refer to the time that the sun spends above the horizon and the time that it spends below it. Both periods have a mean duration of 12 hours. For the Babylonians, a full day—a day and a night—lasted from sunrise to sunrise; for other cultures,

45 Sunrise and sunset
determine the rhythm of
the day.

such as the Greeks, Celts and Jews, it lasted from sunset to sunset. Day
and night were each divided into 12 units, whose duration varied with the
time of year. It is a relatively modern idea to agree on a 24-hour day, starting at midnight, with 60 minutes to every hour and 60 seconds to every
minute: a total of $24 \cdot 60 \cdot 60 = 86,400$ seconds.

Second, there are the phases of the moon. These owe their existence to
the variable positions of the moon in relation to the earth and the sun. The
moon, of course, is not itself a source of light; rather, it reflects the light of
the sun. When the moon lies between the earth and the sun, what we see
from the earth is the side that is not illuminated—the moon is said to be *in
conjunction* with the sun. We call this phase the *New Moon*.[28] As the moon
shifts to positions east of the sun, we see more and more of its right side,
in the phases we call the *Waxing Moon*. At last, the whole moon is bathed
in sunlight—the earth lies between sun and moon, and sun and moon are
in opposition, a phase known as the *Full Moon*. After this, the moon shifts
west of the sun and enters its *waning* phase: darkness creeps across the
moon's face from the right until there is no illumination at all, the *New
Moon* phase again. The period from New Moon to New Moon lasts just a
little bit more than 29½ days[29] and is known as a *synodic* month or, more
popularly, a *lunar* month.

The interval separating the principal phases of the moon (New Moon,
First Quarter, Full Moon, and Last Quarter) is roughly equal to seven days.
This is presumably the reason that the Babylonians and other highly devel-
oped civilizations introduced a further unit of time: the seven-day *week*.
Seven is also the number of planets—in the literal sense of heavenly *wan-
derers*—known to the ancients: Sun, Moon, Mercury, Venus, Mars, Jupiter

46 The phases of the moon are the result of the sun's illumination of the moon and how this is seen from earth.

and Saturn. The seven days of the week were named after these planets by the Babylonians, the Egyptians and the Romans.

Third, there is the rhythm of days getting successively longer or shorter as we pass through the seasons of spring, summer, fall and winter. Babylonian astronomers were the first to notice the great circle that the sun appears to make among the fixed stars in the course of every year. Its path, the so-called *ecliptic*, sees it sweep past the twelve signs of the Zodiac, Aries, Taurus, Gemini, Cancer, Leo, Virgo, Libra, Scorpio, Sagittarius, Capricorn, Aquarius and Pisces.

Today we know that the succession of seasons is caused by the earth's axis forming an angle with the plane in which it moves around the sun. If we were to draw a perpendicular from the earth's center to its orbital plane around the sun, it would make an angle of roughly 23 degrees with the earth's axis. As the earth orbits around the sun, the sun's light is shared differently between the Northern and Southern Hemispheres.

During the spring and fall equinoxes,[30] the Northern and Southern Hemispheres receive the same amount of light, and a line connecting the center of the earth to the sun would pass through the equator. On the day of the summer solstice, the Northern Hemisphere receives the most sunlight, and a line connecting the center of the earth to the sun would pass through 23° northern latitude. On the day of the winter solstice, the Southern Hemisphere gets the largest share of sunlight, and the line connecting the center of the earth to the sun passes through 23° southern latitude. The time that elapses between solstices of the same season, be it winter or summer, is what we call a year—a *solar, tropical* or *astronomical* year—of 365 days and 6 hours.

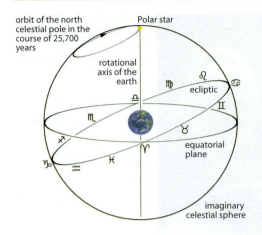

orbit of the north
celestial pole in the
course of 25,700
years

Polar star

rotational
axis of the
earth

ecliptic

equatorial
plane

imaginary
celestial sphere

47 Apparent movement of the sun
against the backdrop of the stars in
the course of a year. The celestial
equator is on the same plane as
earth's, whereas the ecliptic, the
tilted plane of the sun's apparent
path, forms an angle of 23° with the
equatorial plane. In the course of the
year, the sun appears to be moving
through the 12 constellations of the
zodiac that are arranged along the
ecliptic.

The basis of the Babylonian calendar was the rhythm of the lunar
phases.[31] A Babylonian month therefore lasted slightly *longer* than 29½
days. The Babylonian lunar year dealt with this by containing 12 months
with alternately 30 and 29 days. A full year therefore had only (6 · 30) +
(6 · 29) = 354 days. Various systems for bridging the gap between this lunar
year and the solar year of 365¼ days were tried, but by the late fourth century BC, the Babylonians had settled down to intercalating 7 leap months
of 30 days each within every 19-year period. Their reason for choosing
these numbers can be seen from the following equations:

$$(354 \cdot 19) + (30 \cdot 7) = 6{,}936 = 19 \text{ lunar years plus 7 intercalary months,}$$

$$(365 + \tfrac{1}{4}) \cdot 19 = 6{,}939 + \tfrac{3}{4} = 19 \text{ solar years.}$$

This amounts to a disparity of just 3¾ days in every 19 years or about 4
hours and 53 minutes per year.

The Babylonians devised a complex rhythm of their own that synchronized the two rhythms of the lunar phases and the seasons. On completion
of the 19-year cycle, the moon and the calendar agreed, just as they had at
the beginning of this cycle.

For the Egyptians, depending as they did on the yearly Nile floods, the
solar year was the most important temporal yardstick. The Egyptians were
forewarned of the floods by the appearance in the morning sky of one of
the brightest stars in the firmament: the Dog Star, Sirius, in the constellation of Canis Major. To this day we speak of the *dog days* between early July
and August 11, weeks that were decisive for the irrigation of Egypt's fertile
lands with the waters of the Nile.

With such dependence on an *annual* event, it is not surprising that the
Egyptians opted very early on for the 365-day year—called the *Egyptian*

year to this day. Egyptian astronomers knew
very well that their year was short by a quarter
of a day and that their calendar fell behind the
solar calendar by one day every four years.[32]

This problem was eventually addressed by
Julius Caesar. In 63 BC, Caesar was elected
pontifex maximus (chief of the college of priests
responsible for the state religion), a lifetime post
he held until his assassination. The college was
also the guardian of the calendar, and Caesar
used this aspect of his position to introduce a
calendar reform: the *Julian* calendar. Given the

48 Julius Caesar.

great sophistication of this calendar, it is hardly surprising to find that the
college of priests took the advice of the Egyptian astronomer Sosigenes.

One result of the innovations initiated by Sosigenes was to ensure an
auspicious starting date for the new calendar: January 1, 45 BC, which
coincided with the first New Moon after the winter solstice. The main
points of the reform can be summarized as follows:

> The mean duration of a year was set at 365¼ days.
>
> To effect this, a calendar year was set at 365 days a year with an extra
> day intercalated every fourth year.
>
> The distribution of days per month was determined without reference
> to the rhythm of the lunar phases. In leap years, months alternated
> between 31 and 30 days, beginning with January; in ordinary years,
> February lost a day and was left with 29.[33]

Misunderstandings and incompetence within the college of priests
created problems with the new calendar after Caesar's death. Augustus
became *pontifex maximus* in 12 BC, and by 8 BC had recalibrated the calendar according to Caesar's intentions. Some scholars say that Augustus
complicated Caesar's original design by insisting that August (formerly
Sextilis and renamed in his honor) should have the same number of days
as July (known as Quinctilis until it was renamed for Julius). This then led
to the months of September to December being altered to the lengths we
use today to preserve the original 31/30 alternation as closely as possible.
Most scholars, however, now dismiss this story, arguing that Caesar himself shortened February and lengthened August (Sextilis) to accommodate
certain established rituals of the state religion.

People do grow very fond of the rhythm codified in their calendars,
usually bowing to change only after it has become inevitable, as the following examples illustrate.

49 Regiomontanus.

GREGORIVS·XIII· PAPA · BONONIENSIS·

50 Pope Gregory XIII.

In the first example, astronomers at the time of the Council of Nicaea in AD 325 noted that the spring equinoxes no longer occurred on March 24, as they had in Augustus' time, but had moved forward to March 21. The Julian calendar was not perfect, and its imperfections had been increasing at the rate of one day every 129 years. Yet it was not until 1474 that Pope[34] Sixtus IV asked Johannes Müller von Königsberg, a.k.a. Regiomontanus (1436–1476), the eminent Viennese mathematician, to recalibrate the Julian calendar. Regiomontanus' unexpected death only two years later delayed the project by yet another 100 years. Finally, by the time of Gregory XIII's papacy, the beginning of spring had moved forward to March 11th, so Gregory had to rehabilitate the calendar. Taking the Council of Nicaea as his reference point and with the aid of the Italian mathematician Luigi Lilio, Gregory brought the equinox back to March 21 by ordering ten days to be omitted from the calendar of 1582 and decreed that October 15 would immediately follow October 4. Next, he had to ensure that the calendar would not get out of line again. This he achieved by decreeing that the rule for leap years should be modified. Previously, all years divisible by four had been leap years. In the future, the last year of every century would not be a leap year unless it was a multiple of 400. This meant that 1600 was a leap year, but 1700, 1800 and 1900 were not; 2000 was a leap year, but 2100 will not be, etc. In this way, every 400-year period comprises not the Julian $(365 \cdot 400) + 100 = 146,100$ days but rather $(365 \cdot 400) + 100 - 3 = 146,097$ days, which is the closest whole-day approximation to the actual number of days in 400 solar years.

It was a simple enough reform—easy to understand and implement—that actually corrected the problem. Yet in spite of these obvious advantages, the Gregorian calendar was very slow to be accepted. On the date in 1582 agreed upon for its adoption, it became operational only in Italy, Spain and Portugal. France and the Catholic parts of the Netherlands followed toward the end of the year; the Catholic German-speaking countries, Austria and the Catholic cantons of Switzerland, in 1583. Protestant Europe dug its heels in; having rejected the authority of the Pope in matters of religion, it was not about to accept his jurisdiction over time. It was not going to be dictated to by any Pope—the very idea was little less than blasphemy! It was 1699 before the evangelical Diets of the German states accepted the new calendar. The *improved calendar*, as it was called to save face, was ushered into Protestant Germany by the elimination of ten days in February 1700, which meant that February 18 was immediately followed by March 1. In 1701, the majority of Protestant cantons of Switzerland accepted the new calendar, but in Glarus, Appenzell and in a region of Grisons, Protestants continued to use the Julian calendar until 1789. Meanwhile, Great Britain had at last introduced the Gregorian calendar in 1752, with Sweden following a year later. Orthodox Russia continued with the Julian calendar until 1918, which is why the Bolshevik putsch of 1917 is known as the October Revolution, even though, according to the Gregorian calendar, it took place in November.

The second example concerns an attempt at calendar reform that failed, although it would have been truly revolutionary in its simplicity and clarity. Three years after the French Revolution, the Convent put forward the following new calendar:

> The year was divided into 12 months with 30 days each.[35]

> These were grouped into seasons: fall, winter, spring and summer.

> The months were renamed, with names appropriate to the parts of the year in which they occurred.[36]

> At the end of the twelfth month, *Fructidor*, five days were intercalated, the *jours sansculottides* or *complémentaires*. This number was increased to six in leap years.

> Each month was subdivided into three decades of ten days each. In this way, the decimal system would simplify time, just as it had already simplified the old, unwieldy measurements for calculating dimensions, areas, volumes and weights.

Although the decimal system had its way in most other fields, the Republican calendar proved short-lived. Its stark opposition to the Catholic year, deliberately introduced by the free-thinking Convent as a high-profile feature, severely reduced its acceptability. After less than 14 years, Napoleon's decree dated September 9, 1805—rather than 22 Fructidor XIII—put an end to this particular reform. From January 1, 1806 onward, French civic life once more followed the well-trodden Gregorian path.

The third example is a more recent proposal of a reformed calendar dating from the twentieth century. It was hoped at the time that the United Nations would help implement it, but the project never got off the ground. The aim of that reform was more modest in that it did not envisage a complete revision of the calendar—as had the French example—but merely an adjustment to regularize the seven days of the week. It would have made it possible to tell immediately what day of the week any given date was, irrespective of the year. Two trivial alterations were called for:

> December 31 was to have special status. It would not be allocated a day of the week, so that the week in which it fell would actually comprise eight 24-hour periods. December 31 could then be dubbed Hogmanay, Sylvester or whatever name took the fancy of individual countries.

> In leap years, a similar arrangement would operate for a day added at the end of June. Midsummer's Day was considered as a general name for the extra day.

These arrangements would leave every year with 364 ordinary weekdays. As 364 is divisible by 7, the calendar dates of the weekdays would not change from year to year. It would be possible to fine-tune this calendar by introducing the additional proviso that the months of January, April, July and October—the first months of their respective quarters—would have 31 days and the remaining eight, 30.

As 364 is also divisible by 4, one could, by making sure that the first year of the new system began with a Sunday, gain the additional advantage that weekdays were repeated periodically in all four quarters and that all months had the same number of working days: 26. The 31-day months would have five Sundays; the other ones, four. From the point of view of the economy, this calendar would have been ideal.

Harmless though this proposal may sound, it has no chance of being realized in the foreseeable future. This is understandable. Although they eventually accepted calendar reforms, highly developed civilizations have always clung tenaciously to the same old weekday merry-go-round of Sunday, Monday, Tuesday, Wednesday, Thursday, Friday and Saturday—for a day to fall outside this system has so far been unthinkable. Among all the rhythms based on celestial phenomena that structure the passing of time, the weekday rhythm has proved virtually inalienable.[37]

With the invention of clocks, an attempt was made to replicate on a small scale the rhythms of the cosmic movements of earth, moon and sun. The sundial is the most eloquent example. The corollary of the clock-making project is that the rhythms realized in our most advanced clocks are better suited as definitions of time units than the original rhythms traced out on the firmament.

Even before the invention of the pendulum clock, astronomers used the oscillations of pendula of suitable lengths to measure the duration of celestial phenomena. Christiaan Huygens (1629–1695) is believed to have invented the pendulum clock. Galileo[38] had written to Huygens mentioning the possibility of measuring time with mechanical oscillations, and in 1656, Huygens was the first to use a pendulum as the regulating device of a clock. Inventions of ever-increasing sophistication took into account that the cycle of a pendulum depends on such variables as altitude above sea level and ambient temperature, which causes the length of the pendulum to change. At the same time, clockmakers began replacing the long, suspended pendulum by the wheel balance (i.e., a rotating pendulum), especially in pocket watches and table clocks.

51 A clock is a miniature solar system.

Today, quartz timepieces use the electrically incited oscillations of a quartz crystal, which usually comes in the shape of a tuning fork. Provided the temperature of these crystals remains constant, they are practically immune to external influences. Unfortunately, the rhythm with

52 Christiaan Huygens.

which the crystal oscillates does not remain entirely constant; it shows vestigial signs of fatigue after a given time. Because of this, quartz timepieces must be synchronized from time to time by means of a molecular chronograph, which is based on the more precise oscillations of ammonia molecules, or an atomic clock, which pulses to the oscillations of an isotope of cesium. Atomic clocks have attained such a level of exactness that they have even bested the celestial phenomena when it comes to defining time units.[39] Whether an atomic clock is *correct* can only be determined through a comparison with its peers. Earth, moon and sun have now been reduced to the roles of bit players.

Clocks have therefore brought the rhythm of time from inaccessible celestial spheres down to earth. By definition, a day lasts 86,400 seconds—regardless of whether two successive apexes of the sun take this long to occur or not. Similarly, a year lasts 31,556,925.9747 seconds—no matter whether the earth has managed to complete its revolution around the sun in that time or not.

January April July October	February May August November	March June September December
1. Sunday	1. Wednesday	1. Friday
2. Monday	2. Thursday	2. Saturday
3. Tuesday	3. Friday	3. Sunday
4. Wednesday	4. Saturday	4. Monday
5. Thursday	5. Sunday	5. Tuesday
6. Friday	6. Monday	6. Wednesday
7. Saturday	7. Tuesday	7. Thursday
8. Sunday	8. Wednesday	8. Friday
9. Monday	9. Thursday	9. Saturday
10. Tuesday	10. Friday	10. Sunday
11. Wednesday	11. Saturday	11. Monday
12. Thursday	12. Sunday	12. Tuesday
13. Friday	13. Monday	13. Wednesday
14. Saturday	14. Tuesday	14. Thursday
15. Sunday	15. Wednesday	15. Friday
16. Monday	16. Thursday	16. Saturday
17. Tuesday	17. Friday	17. Sunday
18. Wednesday	18. Saturday	18. Monday
19. Thursday	19. Sunday	19. Tuesday
20. Friday	20. Monday	20. Wednesday
21. Saturday	21. Tuesday	21. Thursday
22. Sunday	22. Wednesday	22. Friday
23. Monday	23. Thursday	23. Saturday
24. Tuesday	24. Friday	24. Sunday
25. Wednesday	25. Saturday	25. Monday
26. Thursday	26. Sunday	26. Tuesday
27. Friday	27. Monday	27. Wednesday
28. Saturday	28. Tuesday	28. Thursday
29. Sunday	29. Wednesday	29. Friday
30. Monday	30. Thursday	30. Saturday
31. Tuesday		+
		31. Sylvester/New Year's Eve/ Hogmanay (in December)
		31. Midsummer's Day (every leap year in June)

$$26 + 5 + 26 + 4 + 26 + 4 = 91 = 364 / 4$$

53 A UN world-calendar year with 364 traditionally named days would make a sensible distribution of the weekdays within each quarter possible.

HOFMANNSTHAL: NUMBERS AND TIME

If you ask a physicist about the nature of time, in nine cases out of ten, you will get the answer that time is what clocks measure. The answer is both correct and incorrect, depending on what is meant by it. It is correct if what is understood by *clock* is a device to indicate the rhythm to which time is subject. It is incorrect if it implies that it is possible to use a clock to shrink time to a geometrical concept.

A few more remarks may be necessary to clarify this point.

A clock specifies time by having its hands indicate a point along a straight line. For it to be able to do that, the straight line is calibrated at the outset. Two points that are not identical are arbitrarily plotted on the line to stand for, say, zero hours and 13:00 hours—the other points of reference on the scale follow naturally from the first two. It is a mere matter of convenience that most clock faces exhibit this straight line rounded to a circle with twelve principal points of reference; our geometrical representation remains unaffected by this. Nor have digital displays, fashionable since the 1970s, made any real difference. It is just that the analog picture of a clock hand moving along a scale is closer to our perception of time than a succession of digits flashing up on a screen.

Basically, a calendar is nothing but a clock face with a different type of scale marked on it. The short intervals are days; the long ones, years.

The mechanism that causes the clock hand to move along the straight line differs from clock to clock. In the case of the sundial, the shadow cast by the gnomon is propelled by the apparent movement of the sun. The pendulum in a grandfather clock is kept in motion by the earth's gravity; a balance, by the elasticity of a coil; a quartz crystal, by the electric current provided by a battery. All clocks, regardless of what makes them tick, are designed to measure something that appears to exist independently of them. This is as true of the pendulum clock at Hamburg airport, the 99.436-centimeter pendulum of which completes exactly 86,400 cycles during one complete revolution of the earth about its axis, as it is of America's largest water clock in the Children's Museum in Indianapolis. Neither clock *knows* the extraordinary feat it is performing. This quality of clocks led Isaac Newton (1643–1727) to formulate what he considered was one of the most far-reaching insights he had gained: the discovery of *absolute time*. Absolute time is the key notion underlying the theory that all physical phenomena, from the movements of the heavenly bodies to the oscillations of the atoms, partake of *one* grand cosmic context, rest on *one* unifying basis and, above all else, evolve along *one* time scale.

Supplementing rather than contradicting Newton, Albert Einstein (1879–1955) demonstrated that unifying basis to be the reason why every clock measures a time that is uniquely and specifically its own. He also

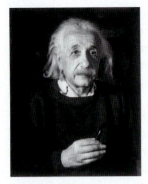

54 Albert Einstein.

taught us how these specific times can be related to each other by taking into account both the relations between individual clocks and the relations between these clocks and their cosmic environment. A clock that changes its location relative to other clocks will be slower than an immobile one. So it might well be literally true that traveling preserves one's youth—though the effort involved and the pitiful fractions of nanoseconds theoretically gained will have few people rushing to try the experiment! When a clock is in the vicinity of very dense matter or even, in the most extreme case, a black hole, it will slow down dramatically—to be in the presence of matter of any kind spreads the time scale. These observations are without doubt among the most fascinating aspects of Einstein's Theory of Relativity.

More disquieting than all of this is the realization that there is no such thing as an ideal mechanism for clocks. Every one of them is doomed from the outset to ultimate disintegration. Even the rotation of the earth about its own axis provides no absolutely reliable benchmark. A day is supposed to have the same duration as one complete revolution of the earth on its own axis. Reckoning in the millions of years appropriate to geological developments, we will discover that, as every century elapses, the *day* lengthens by 0.00164 seconds. Since the dawn of humankind, this has added up to something like 20 seconds—an increase which may appear negligible, yet is nevertheless indisputably there. One of the reasons for

55 A moving clock runs slower than a stationary one.

this stealthy lengthening of the day is the moon, the principal shaker and mover of ebb and tide. The tidal swelling of the waters acts a bit like a brake parachute as the earth rotates about its own axis. It is a fact that the rotation of the earth about its axis, which determines the length of a day, and the revolution of the earth around the sun, which determines the length a year, are asynchronous.

Why clocks run at different speeds depending on whether they are in motion or standing still, how clocks behave in empty space and in the magnetic fields of enormous masses of matter are questions, as has already been noted, for the Theory of Relativity to sort out. For an explanation of why all clocks eventually go lame and why the same is true of all processes, no matter how regular and periodical they appear initially, we must turn to thermodynamics, a physical theory inseparably associated, at least in German-speaking countries, with the name of Ludwig Boltzmann (1844–1906). According to one of the fundamental theorems of thermodynamics, all concentrated energy will ultimately dissipate. A wound-up clock, for instance, is a system charged with concentrated energy, Leibniz' *vis viva* (Latin for "living force"). Yet given sufficient time, the coil will relax, and

56 Ludwig Boltzmann.

the energy invested in it will be passed on to the environment as diffuse heat. The solar system behaves no differently—except in the time scale involved. It takes eons for the concentrated energy stored in the solar system to be dissipated into the cosmos.

Just as the rhythms of pendula and of our own pulse ebb away, so will that of constellations, eventually. This is why the image of time atrophied to a calibrated straight line is so inadequate. According to Einstein, it is not a straight line at all but a tangle of warped and twisted curves. The one exception is *eigentime* (personal time), which is experienced as linear by the observer. According to Boltzmann, it is straight only for a little way before, so to speak, "losing its sense of direction". Above all, it is this geometrical metaphor that prevents us from understanding the *passing* of time. For us, a straight line is something that is simply *there* in front of us; in the same way, the time scale is spread out where we can see it. At least in thought, we can select any spot and move forward and backward on it. Yet the fact remains that we can truly experience only the present instant, which moves implacably forward on the scale, and lack even the capacity to make it stand still. Why is this so?

This experience of how we are time's hostages reminds us of the sense of bafflement overwhelming Hugo von Hofmannsthal's Marschallin: "Such a mystery, / Such a very great mystery."

In a genuinely mathematical attempt to get some perspective on this mystery, let us try reading the calibrations on the straight time scale as numbers. To be more precise, let us find out what happens when we, inevitable losers in time's race, experience numbers, over which time has no sway.

We listen to a clock striking in a tower: one, two, three, four, five, six, seven, followed by a silence. *Seven* is the number that the clock is using to tell us the time. We understand this on the basis of our counting skill.

Pursuing this seemingly trivial insight further, we may note that the seventh stroke sounded no different from, say, the first or the fourth. However, as part of counting *up to* seven, we realize that we have already gone past one and four in counting, that the first and fourth strokes have already receded into the past. The seventh stroke was the last one—this need not have been the case. We know that seven could be followed by eight, that, generally speaking, no numeral is ever the last word. We do not stop at seven out of sheer whim; on the contrary, we listen to find out whether the clock will continue striking, forcing us thereby to continue counting. In listening, we actualize our anticipation of the future.

It is by counting that we conclude that numbers are not, as the simplified image of the straight scale with its calibrated marks would suggest, a commonplace commodity entirely at our disposal. As we pronounce a number, we subsume mentally under that number *all* of the numbers that precede it. We begin with *one,* and we know that whatever number we have just pronounced can *never* be the last number of all. Let it announce itself with all the swagger of Leporello's *mille tre* at the beginning of Mozart's *Don Giovanni*; there will always be another to follow on its heels.

The same is true of time: as we consciously experience it, we become aware that the present moment does not stand isolated by itself. It follows naturally from the sequence of past moments, and it will not be the last one. It is therefore irrelevant whether we measure the flow of time with the rhythm of our own pulse, the oscillations of a pendulum or the rising of a star in the sky. It is in counting—the foundation of all rhythms, the basis for all kinds of periodicity—that time manifests itself. This above all else is why time must be an intrinsically human construct—a construct that we humans have only projected onto the cosmos.

And here we come to the innermost core that numbers and time have in common. Both of them appear to us as externally given, and we experience them as if they were legitimate objects for our outward-bound senses.

57 *Tempus fugit.* Language lends itself all too easily to the dictum of how time flies. Yet it is not time that passes, but ourselves. Our lives pass. We change identities as mountebanks change their impersonations. So it is with numbers: it is not numbers that do the counting, we ourselves are the tellers—and the tellers of tales.

In fact, however, neither of them is a part of the natural order of things. There is no place in the entire universe where numbers can be found independent of us; both of them are *rational* constructs, creatures of our thought and of our consciousness.

The fact remains, however, that in our heart of hearts, we experience the irreversible flow of time as something profoundly painful. Can this have something to do with the fact that we experience *numbers* as entities beyond our control, radically outside our sphere of influence? Measuring time with an enormous variety of different rhythms, from our heartbeat to the cosmic periods of earth, sun and moon in the firmament, we have become used to an equally enormous variety of methods of counting. Yet how we count has as little to do with the actual existence of numbers as how we measure it has to do with the actual existence of time.

Of course we can decide to break off at seven—yet we know that eight looms ahead with the same intransigence that characterizes the clock hand as it moves on.

It is in our power to count regularly or irregularly, quickly or slowly and so on, but this means only that we have opted for our personal style of counting. We have, as it were, gauged our time scale according to our individual taste. Once we realize that the *same* process of counting can be actualized in a multitude of *different* ways, more quickly or more regu-

58 What we celebrate with the fireworks of New Year's Eve is starting anew: we turn over a new leaf and start counting from scratch.

larly, more slowly or haltingly, we are free to compare different rhythms and to agree on one standard. In antiquity, this standard was found in cosmic events; today, physicists generate it in their labs. This has no effect as such on the inexorable procession of numbers, and numbers are our only real basis for understanding time.

It is also true that we can arbitrarily decide, as we count, to switch back to *one* and to start again, from scratch. This does not mean that we have reinvented counting; we have merely rebooted the original file, as it were, and started again. The same applies to time. It is, of course, possible to reset the starting point of time at any moment, and there is a sense in which we do this, every one of us, with each conscious opening of our eyes. We also do this as a community. The fact that the scale on our timepieces has been assigned the shape of a circle means that we can start afresh with *one* after half a day has passed. The euphoria during the night of December 31/January 1 is due to the collective illusion that, by winding back the process of counting to *one,* we can ring in a better era with the New Year.

On a more comprehensive scale, chronology attempts to tie what we count from to some historical event of superlative significance. For the Romans, the presumed year of the foundation of their capital marked the beginning of their chronology. Modern western civilization regards the birth of Christ as calculated by Denis the Little, a sixth-century Roman monk, as such an event.[40] Needless to say, Muslim, Chinese and other civilizations have their own points of departure for the reckoning of their eras. In all of this, we tend to overlook the fact that setting the beginning of an era presupposes a reference point in another era. The liberty to propose a new beginning is always tied to memory, which encapsulates and preserves the past, and to a procedure of counting that was already in place before that new beginning.

It would be possible to suggest a chronology that derives from some *absolute* event. In our science-dominated civilization, an obvious candidate for such an absolute event would be the *Big Bang*, in which the whole universe burst forth from the infinitesimal point that we call *singularity.* However, it is not only practical reasons that militate against the Big Bang being used to mark the beginning of Year 1. Among other things, it is hopeless to trace back the timeline to the Big Bang with anything approaching certainty. We do not even know the exact number of billions of years by which this event antedates us. Another reason is the fact that the Big Bang as such is completely without meaning. It is referred to only in the context of astronomers attempting to construe something resembling a *collective memory.* The existence of the Big Bang is restricted to its role as a theoretical construct. To that extent, it is no more suitable as a reference point for our reckoning of time than is any other event chosen for that purpose at random—there is no shortage of more plausible candidates.

There is an orthodox Jewish tradition according to which God created the world 3,760 years before the beginning of the Christian era. Within this tradition, this is when time reckoning in "absolute" terms has to start. This seems no less (and no more) legitimate than postulating a beginning of absolute time, on the basis of scientific cosmological hypotheses, some 13, 14 or even 15 billion years ago.[41]

One final question remains to be asked: since time is more intimately bound up with our individual destinies than anything else, what permits us to speak meaningfully of *time* in the singular? Given that time and numbers are inextricably interwoven, this question can be rephrased as, "Why are the same numbers *equally valid* for everybody? How is it that no doubts ever arise as to how to count *in principle*, irrespective of speed, regularity and when to begin?"

Leibniz provided one possible answer to this question with his disturbingly far-fetched notion of "pre-established harmony". His argument can be summarized as follows: although it is true that we are all of us windowless monads, individuals unable to interact with one another, mere self-referential systems, it is equally true that God has created the world in such a way that the rhythms of all such monads are synchronous, like thousands of clocks built by the hands of one and the same clockmaker. It is on this coincidence that we have built the illusion that there can be any true communication between individual souls.

Such is the unedifying worldview advanced by Leibniz. More convincing and certainly more heartening is that propounded by Erwin Schrödinger (1887–1961). According to Schrödinger, individual consciousnesses are merely different realizations of a *single* underlying general consciousness, in the same way that all of the dazzling rays of light reflected or refracted by a diamond can ultimately be traced to a single source. This single consciousness is

59 Erwin Schrödinger.

60 The birth of Christ (as defined by
Dennis the Little) marks the beginning of
the Christian era.

61 How to locate the *absolute* beginning of
counting?

the source of our shared experience of time. Being founded in an innate
understanding of numbers which seems to emanate from those strata of
consciousness that are common to us all, it has nothing to with the individual.

Perhaps this insight of Schrödinger's can help us understand the last
words of Hofmannsthal's *terza rima* on transience, in which he acknowledges with a sense of shock,

> And kinsmen laid in coffins long ago
> Are yet as close to me as my own hair,
>
> Are no less one with me than my own hair.

Descartes

Numbers and Space

"Why are the mountains blue?" Like many another little child, young Giordano Bruno is said to have been pacified by his father with the answer that this was because they were such an awfully long way away.

One is tempted to think that, in later life, this shocked awareness of the immensity of space was never too far from Bruno's mind. It was decisive for his thought; he kept it alive with all of its elemental force and attempted to influence others to share his vision. Even before he was able to support

René Descartes.

63 Giordano Bruno.

his convictions with astronomical data, he was the first to assert that the sun is just another star among countless others, all projecting their light to us from the farthest reaches of outer space, from a distance so great that their blazing light has dwindled to little more than pinpricks in the sky.

Before Bruno came along, the universe had been regarded as a compact astronomical sphere. God stood *outside* it, holding it in his hand. Bruno saw the universe as an entity without limit or boundaries, without an outside where God could stand. The only place Bruno could find for God was *inside* things, inside the dust particles, the stones, the stars. The notion of the infinity of space made Bruno embrace pantheism, which had been condemned by the Church as a heresy. Unable to find a place for Bruno's theories within its teachings, the Church resorted to excommunicating him. Eventually, on February 17, 1600, he was burned at the stake.

Obviously, it is sometimes dangerous to go out on a limb where thinking and talking about space are concerned. This is astonishing—especially because the experience of space can justly be considered to have been one of the very first scientific experiences.

Geometry, our name for the body of mathematics that concerns itself with the properties of points, lines, surfaces and solids, is essentially a Greek word. It translates literally as *terrestrial surveying; ge* is *earth* or *land* as in *ge*ography and *ge*ology, and *metrein* is the verb *to measure*.

The beginnings of geometry probably date back to the practice of surveying in the first developed civilizations and in the prehistoric periods of those peoples who built Stonehenge, Glenquickan and other megalithic stone circles.

For the Egyptians, a knowledge of geometry was vitally important in reapportioning land to farmers after the annual Nile floods. The ability to construe a right angle was indispensable for staking out rectangular fields. How they did this is not entirely clear, but they probably took a length of rope, tied 12 knots into it at regular intervals

64 Egyptian official surveyors.

and then tied up the two ends. They then laid out the rope in the shape of a triangle in such a way that the sides had three, four and five interstices between knots, respectively. The resulting triangle has an exact right angle opposite its longest side.

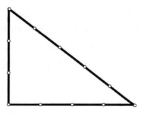

65 A right-angled triangle.

We do not know whether the Egyptians were aware that this construction yielded an exact right angle. Maybe they had a shrewd idea that it did, or maybe they felt the construction gave them a right angle that was sufficiently precise for their surveying purposes.[42] We do know, however, that both Babylonian and Greek mathematicians were able to prove that a triangle with sides of three, four and five units, respectively, is an exact right triangle.

Bhaskara, a scholar in the twelfth century AD, whose work is a veritable treasure house of Indian mathematical knowledge from its earliest beginnings, conducts the proof as follows. Bhaskara considers four right triangles in each of which the right angle is enclosed by sides of three and four centimeters.[43] A pair of these, joined together at their longest side, forms an indisputable rectangle with an area of 3 · 4 = 12 square centimeters. As will also immediately be seen, the two acute angles add up to one right angle. He now joins these four right triangles to form a square in such a way that the hypotenuses of the triangles serve as its sides.

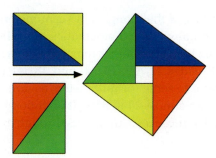

The resulting figure displays a slightly rotated square in the middle, whose sides measure 4 − 3 = 1 centimeter. Bhaskara's laconic comment on his self-devised figure is, "Behold!"

The task he sets for his students is to work out the area of the big square. This is done by adding up the area of the small square of one square centimeter and the area of the 2 · 2 = 4 triangles of 2 · 12 = 24 square centimeters. Because the area in question is 1 + 24 = 25 = 5 · 5 square centimeters, it follows that the third side of the right triangle is 5 centimeters long.[44] The

same is true of a right-angled triangle whose right angle is enclosed by
sides 8 and 15 centimeters long.

Bhaskara's figure enables us to calculate the length of the third side. All
we have to do is to add up the following numbers:

$$(15 - 8) \cdot (15 - 8) = 7 \cdot 7 = 49,$$

which is the area of the slightly rotated square in the middle, and

$$2 \cdot (15 \cdot 8) = 240,$$

which is the area of the four triangles.

A pair of these triangles, as has already been demonstrated, can be
joined together to form a rectangle with an area of $15 \cdot 8 = 120$ square centimeters. Because the sum of $49 + 240 = 289 = 17 \cdot 17$, it follows that the third
side of the triangle is 17 centimeters long.

One more example for the same configuration is a right triangle whose
right angle is enclosed by sides 5 and 12 centimeters long. The arithmetic
is as follows:

$$(12 - 5) \cdot (12 - 5) = 7 \cdot 7 = 49$$

$$2 \cdot (12 \cdot 5) = 120$$

$$49 + 120 = 169$$

$$169 = 13 \cdot 13.$$

Therefore, the third side of the triangle is 13 centimeters long.

Tradition says that Babylonian mathematicians of the time of Hammu-
rabi (c. 1792–1750 BC) knew of another member of this select club: a right-
angled triangle with sides 12,709, 13,500 and 18,541 units long, respectively. Even if we were to settle for millimeters as units, a drawing of this
triangle would take up almost 20 meters. It is impossible for the Babylonians to have arrived at these measurements through experiment, because

66 In a right-angled triangle
where sides of 1 and 3 units
enclose the right angle, the
square of the length of the
third side would have to be
10—and the square root of
10 is *not* an integer.

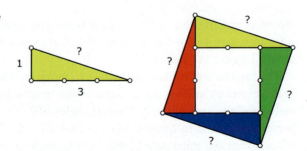

they would not have had the necessary draftsmanship at their disposal.[45] We have no way of knowing how the Babylonians arrived at the extraordinary specifications for this gigantic right triangle.

If, in specifying such a triangle, we were to try out lengths for the sides enclosing the right angle which have been chosen at random, such as 1 and 3, it would follow from the Bhaskara calculation, for which the addends are

$$(3 - 1) \cdot (3 - 1) = 2 \cdot 2 = 4$$

and

$$2 \cdot (3 \cdot 1) = 6,$$

that the area of the big square is $4 + 6 = 10$ and the length of the third side of the triangle is equal to the square root of ten. This raises the problem that ten is not a square number—there is no whole-number root of ten, no factor which when multiplied by itself yields ten as the product. Only certain triads of numbers, such as (3, 4, 5), (8, 15, 17), (5, 12, 13) or indeed the triad discovered by the Babylonians (12,709, 13,500 and 18,541), permit the construction of such right-angled triangles as just discussed.

We must assume that the Babylonians had discovered a method of arriving at such triads, but we have no clues about what that method looked like. What we do know is how the Pythagoreans (and, in their wake, Plato) were able to calculate them systematically.[46]

How triangles behave whose side lengths cannot be expressed in whole-number multiples of any given unit of measurement will be discussed later. For the moment, we will just consider some of the conclusions to be drawn from Bhaskara's tremendous intellectual achievement.[47]

Over the centuries of human civilization, surveyors have engaged in systematic study of the properties of right-angled triangles and the possible ways of calculating their sides. However, even in their most sophisticated deliberations, they have never strayed far from Bhaskara's figure as described earlier. The following is a typical task from a Babylonian book of geometry problems: "A 30-ell beam leaning against a vertical wall has slid down so that its point of contact with the wall is now six ells lower than it was. What is the distance that now separates its point of contact with the ground from the wall?"[48]

A clue follows immediately: "6 deducted from 30: behold, 24." The

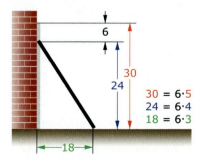

67 A task from a Babylonian mathematics textbook.

author of the problem refers to the right-angled triangle, whose measurements are 30 ells (the length of the beam) for the hypotenuse and 24 ells (the vertical distance between the beam's top end and the ground) as one of the adjacent sides. Of course, $30 = 6 \cdot 5$ and $24 = 6 \cdot 4$. We have already seen that it is possible to construct a right-angled triangle using 3, 4 and 5. If the unit of length measurement is defined as 6 ells, it follows that the third side of the triangle is $6 \cdot 3 = 18$ ells long. This is indeed the solution, as the Babylonian scholar explicitly tells us, "The distance it has covered on the ground is 18 ells." To make sure that all of this is clear, he rather pedantically reverses the question at the end: "If the beam has covered a distance of 18 ells on the ground, how far has it moved downward on the wall?"

Another problem centers on the equilateral triangle. We can easily demonstrate that each of its internal angles is two-thirds the size of a right angle. Using the commonly accepted basis of 90° for a right angle,[49] the internal angles of an equilateral triangle therefore measure 60°. To demonstrate this, we draw a line from a vertex to the midpoint of the opposing side, which not only halves the angle in the vertex but also subdivides the whole equilateral triangle into two congruent right-angled triangles. It follows that the internal angle of the equilateral triangle augmented by one half of its value results in a right angle of 90°—and for this to be possible, the internal angle must measure 60°.[50]

The next logical step from this insight is as follows: if one of the acute angles of a right triangle measures one-third of a right angle (i.e., 30°), the

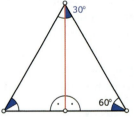

68 An equilateral triangle has three internal angles of 60° each.

69 Surveying instruments based upon Ptolemaic triangulation on the terrace of Belvedere Castle in Prague; these instruments were used by Tycho de Brahe and Johannes Kepler.

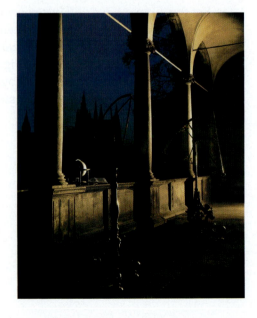

opposing side is half as long as the hypotenuse (the side opposite the right angle). In
mathematical jargon, this is what mathematicians mean when they say they have calculated the *sine*[51] of 30°. The sine of any angle
is the ratio of the side opposite that angle in
a right-angled triangle to the hypotenuse of
that triangle. The sine of 30° is therefore 1:2 =
0.5. Using somewhat more complex reasoning, Claudius Ptolemy, a Greek astronomer
and geographer of the second century AD
and the father of the geocentric Ptolemaeic
system, calculated the lengths of the opposing sides for a wide variety of angles where
the length of the hypotenuse was known.
Yet however intricate Ptolemy's calculations
became,[52] their ultimate foundation is the
same flash of genius that inspired Bhaskara
to draw the figure we just described.

The right-angled triangle was also
instrumental in the first reasonably accurate determination of the size of the earth.
This was carried out by Eratosthenes of
Cyrene around 200 BC and is one of the
most outstanding achievements of science
in antiquity. It is all the more remarkable
because, at the time, the fact that the earth
is a sphere had not even been established
beyond doubt and geographers' knowledge
of its surface was still fragmentary.

Eratosthenes had two points of reference.
He knew that at noon on June 21, the day
of the summer solstice, the sun would be
reflected in the bottom of a deep well in the
city of Syene, meaning the sun was directly
overhead in the sky at this time. He was also
able to find out that the 40-meter-high obelisk at Alexandria, which stands around 800
kilometers[53] north of Syene, casts a shadow
5 meters long at noon on June 21. From this
data, he determined that if two lines could

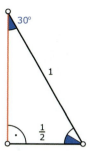

70 The sine of 30° equals 0.5.

71 Claudius Ptolemy holds
a model of his geocentric
universe in his hand.

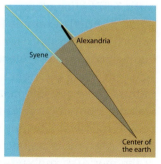

72 Eratosthenes' method for
calculating the circumference
of the earth.

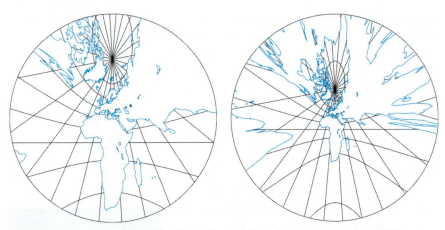

73 If the earth's sphere is projected from the earth's center onto a plane tangential at Syene, the resulting distortions increase the farther we move from the African continent.

be drawn—one continuing the shaft of the Syenean well and another continuing the line of the Alexandrian obelisk—each extending to the center of the earth, they would intersect at an angle of 7.2°. Factoring in the 800-kilometer distance from Alexandria to Syene allowed him to obtain a figure for the circumference of the earth: 40,000 kilometers. He could also obtain a figure for the radius of the earth—800 kilometers (the distance from Syene to Alexandria) divided by R (the radius of the earth) is equal to 5 meters (the length of the Alexandrian obelisk's shadow) divided by 40 meters (the height of the obelisk itself). So

$$R = (40/5) \cdot 800 = 6,400 \text{ kilometers},$$

which is close enough to our currently accepted figure for the radius of the earth, 6,371 kilometers.

Let us briefly examine Eratosthenes' reasoning in greater detail. It was of decisive importance for him that he could be sure that a distance of 800 kilometers separated Syene and Alexandria. This is the shorter of the two sides enclosing the right angle, the other one of course being the earth's radius. Those 800 kilometers had been measured step by step by people walking from one city to the other. Strictly speaking, the line that these people followed was not straight, but curved. It was, at least in theory and if Alexandria had actually been due north of Syene, a short section of the meridian running through the two settlements. In other words, it was an arc. However, Eratosthenes' calculation is not invalidated by this objection. The arc he used is so short compared to the whole length of the meridian that its curvature is negligible. To achieve absolute precision, Eratosthenes would have had to draw a line tangent to the meridian from the well at Syene to the obelisk at Alexandria (or rather its straight-line projection).

The preceding objection may appear finicky at first sight but comes into its own in map-making. If the task is to render parts of the earth's surface cartographically, we are forced to establish a relationship between the spherical shape of the planet and the two-dimensional nature of maps. One option[54] in map-making is to construct a plane tangent at a given pole of projection or map origin—let us say at Syene. Places in the immediate vicinity of Syene will then be projected in a straight line onto the tangent plane from the earth's center. Reduced to scale on a sheet of paper, the tangent plane will appear as a map on which the location of Syene and its environs are shown accurately but which displays rapidly increasing distortions the farther away we move from that center.

This may appear to be a less than ideal cartographic method. True, but if we take a closer look at the problems involved, we will soon realize that there is no such thing as an ideal method for depicting large portions of the earth's surface.

Let us briefly consider a triangle on our planet that has the North Pole as its vertex and the meridians running down to the equator through New Orleans and the Fiji Islands as two of its sides. Their points of intersection with the equator would indicate the triangle's two other angles, with the equatorial arc— equal in length to the meridian arcs—connecting them and thus forming the triangle's third side. Obviously, this three-sided figure would be equilateral, yet it would also have three right angles. Before we could manage to produce a scale map of the planet that rendered all dis-

74 The sum of the internal angles of a spherical triangle exceeds 180°.

tances and angles accurately, we would need some way of drawing a two-dimensional figure with three equal sides and three internal right angles. We are forced to conclude that two-dimensional maps must always distort what actually exists on the surface of the globe.[55] In other words, if we use Bhaskara's figure as our base for calculations of great distances on two-dimensional maps, we will always produce results that differ a great deal from those actually measured on the earth's surface. By comparing the results of calculation and of measurement, we have used *geometry* to prove that the earth cannot possibly be disk-shaped.[56]

75 Pierre de Fermat.

Calculus is a mathematical theory developed by French lawyer Pierre de Fermat (1601–1665), English physicist Sir Isaac Newton and that German universal genius, Gottfried Wilhelm Leibniz. It enables us to construct a level plane tangent at one point on a curved plane and to calculate the angles and distances projected from the curved plane by dint of Bhaskara's figure. In the immediate vicinity of the point of contact, there will be a close correlation between the distances and angles measured on the curved plane and their counterparts calculated on the tangential plane. Yet once again moving away from the point of contact shifts the results on the tangential plane further and further away from what is actually happening on the curved plane.

If we apply calculus to the previous example, the tangential plane touching the earth at Syene, the distortions we encounter will be as follows: Alexandria, in reality 800 kilometers away from Syene, will appear on the tangential plane to be 804 kilometers away, a negligible distortion of only 0.5%. The distance from Syene to the North Pole is almost ten times greater, some 7,460 kilometers; on the map, this distance will appear as approximately 15,026 kilometers. Here the error is close to 100%. Finally, we might consider a line connecting Syene to the north coast of Antarctica. In fact, it covers about 10,000 kilometers; on our hypothetical map, it would appear to stretch out to infinity.

Let us return to Eratosthenes and the first measurements and computations of very long distances. Roughly one century after Eratosthenes' great achievement, the distance between the earth and the moon[57] was calculated to within a few percent. Credit for this extraordinary feat goes to the Greek astronomer Hipparchus (d. after 127 BC), who again based his research both on geometrical calculations—albeit more sophisticated ones than those of Eratosthenes—and on empirical data obtained from the observation of lunar eclipses.[58]

Half a century before Eratosthenes, Aristarchus of Samos (310–230 BC), another Greek astronomer, even attempted to calculate the distance between the earth and the sun. His geometrical method, in which right-angled triangles were again a key concept, was correct in principle but so skewed by imprecise measurements that his result (he calculated that the sun was a mere 19 times farther away from earth than the moon) is a long way from the mark. We now know the sun to be some 400 times farther away from the earth than the moon.

76 Aristarchus and Hipparchus.

Nearly two thousand years later, the famous German astronomer Johannes Kepler (1571–1630) lived out his life without any idea of the real distance between the earth and the sun. What he did establish, applying the laws of planetary motion he had discovered, was that the distance from Venus to the sun was roughly 72% of the distance from the earth to the sun. Using the same methods, he was able to determine that the radius of the orbital path of Mars was about one and a half times that of the earth. He then went on to calculate with the same degree of accuracy the relative distances of all of the other planets from the sun. This enabled him to draw an accurate scale diagram of the solar system. However, no one in Kepler's seventeenth century had any idea at all about absolute distances within the solar system.

77 Johannes Kepler.

It took another hundred years for an exact measurement of the distances within the solar system to become possible. On June 6, 1761, and June 3, 1769, two very rare astronomical events occurred—*transits of Venus*, during which Venus is seen from earth moving across the disk of the sun. As Venus' orbit lies within that of the earth, there must be occasions when the earth, Venus and the sun are aligned roughly along the same axis. Usually, Venus is slightly out of line and therefore passes between the earth and the sun either just above or just below this axis. Every 130 years, however, the three heavenly bodies—sun, Venus and earth—are in perfect alignment, and Venus can be observed from earth for several hours as a black speck making its way across the bright face of the sun. Transits of Venus come in pairs, the second occurrence trailing the first by eight years.[59]

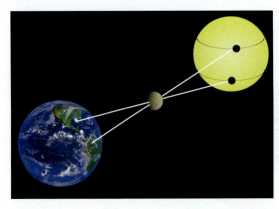

78 During a transit of Venus, Venus passes directly between the earth and the sun; measurements of these angles, taken from two different and sufficiently distant points on earth, make a survey of the solar system possible.

When a transit of Venus is observed from two points a considerable distance apart on a north-south axis, Venus appears to be traveling across the sun along two *different* chords, a phenomenon that the eighteenth-century astronomers knew how to exploit to their advantage. They constructed right triangles extending from their observation posts to Venus and the two chords on the solar disk beyond. They then calculated the ratio of the adjacent sides on the basis of the angles they had obtained from their observations. Precise measurement of the distance separating the observation points completed the array of data needed to calculate the distances of Venus and the sun from the earth in absolute terms.

It is highly instructive to make an accurate scale drawing of the earth and the moon and to get a first-hand impression in this way of the proportions and distances involved. Another bonus is an appreciation of just how truly magnificent an achievement it was for mathematicians to have provided the exact data needed for men to set foot on the moon in 1969. A schematic drawing of the sun-earth system in which 15 centimeters represent the distance between the two bodies shrinks the sun to a proportionate diameter of 1.4 millimeters and the earth to a speck smaller than a grain of dust with a diameter of 0.013 millimeters. In the same drawing, even the lunar orbit is no more than a paltry 0.8 millimeters in diameter.

Yet all this is nothing compared with other distances calculated by astronomers. When Copernicus put forward his heliocentric system in the middle of the sixteenth century, he and his supporters were faced with an enduring

79 Nicolaus Copernicus.

80 Sketch of the Copernican system in Copernicus' own hand and reprinted in his *De revolutionibus orbium coelestium libri sex.*

embarrassment largely due to his premises concerning the nature of the cosmos. Like the astronomers in antiquity, he believed that the fixed stars were brilliant points of light set into a gigantic crystal sphere that contained the whole universe. If it were true that in the course of a year the earth completed an orbit around the sun measuring some 187 million miles, then surely a fixed star would be seen from earth in different positions on a summer night and on a winter night. Yet observing such a difference, known as a parallax, eluded the astronomers of the time.

One of Giordano Bruno's apparently more fanciful ideas, namely that the fixed stars were so immeasurably far away—several thousand times the distance from the earth to the sun—that their parallaxes are pushed below the threshold of observation, offered the only way out of this tight corner for the Copernicans. This, however, did sound very much like an explanation cobbled together in an emergency, and it was the butt of anti-Copernican ridicule for decades.

It was only in the nineteenth century that optical instrument makers became capable of building telescopes accurate enough for the Kaliningrad (then Königsberg) astronomer Friedrich Wilhelm Bessel (1784–1846) to measure the parallax of a fixed star in 1838. The minor star observed by him, one called *61 Cygni* in the constellation Cygnus (the Swan or Northern Cross), does indeed have a barely noticeable shift, the oscillation being no more than approximately one thousandth the visible diameter of the lunar disk. Barely noticeable it may be, but it is enough to prove that the earth goes around the sun.

81 Friedrich Wilhelm Bessel.

But back to distances: using the geometrical methods grounded in Bhaskara's figure, Bessel calculated this star's distance from the earth with a breathtaking result. On October 19, 1838, Bessel wrote to his colleague Wilhelm Olbers (1758–1840) in Berlin: "The parallax of the star '61 Cygni', which I have measured, corresponds to a distance of 98,330,000,000,000 kilometers (more

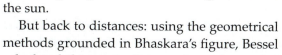

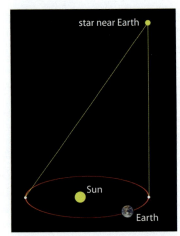

82 If the earth revolves about the sun, light coming from a star to the earth must change its angle of approach in the course of half a year.

83 The galaxy NGC1512 is a system unimaginably distant from earth containing billions of stars.

than 61 trillion miles)[60], a distance which light takes 10.3 years to travel." This innocuous missive caused the supposedly star-studded crystal sphere to shatter and reveal the unfathomable depths of the universe.

Let us now apply our schematic diagrams to 61 Cygni. If we draw the star and our own earth as two dots 20 centimeters apart, we must shrink the earth's orbit around the sun proportionately to 0.0006 millimeters, a quantity far too small to be perceived with the naked eye. Meanwhile, we must show the earth's diameter as a mere 0.00000002 millimeters wide, almost infinitesimally small, no more than a fraction of an atom. Yet 61 Cygni is one of the few stars that are relatively close to our solar system. Most stars even in our own galaxy are several thousand times farther away.[61] And our galaxy is only one of countless other stellar systems, each containing billions of stars, most of which easily outshine our sun. Our galaxy's nearest neighbor in space is the Andromeda galaxy, a close neighbor in cosmic parlance, but the journey there would take two and a half million years to complete, even traveling at the speed of light.[62] Compared with this, the 10.3 light-years' distance of 61 Cygni does not seem so impressive.

Who, among the ranks of angels, would hear me if I screamed?

So starts the "First Duino Elegy" by the Austrian writer Rainer Maria Rilke (1875–1926), with the cry of a lyric poet driven to the brink of despair by a kind of absolute loneliness. Imagine a man staring up into the heavens, gradually becoming aware of space in all its vast emptiness. Imagine him beginning to see his own world soaring—again in Rilke's words— "into loneliness remote from all stars", utterly lost in the gaping desolation of the universe. Might this not be the very question that would burst from his heart?

As early as 1661, the Italian mathematician Vincenzo Viviani (1622–1703) designed an experiment to prove the earth's rotation about its own axis. If we set up a pendulum on the North Pole and set it in motion, the plane of oscillation will appear to rotate a full 360° in 24 hours. During that time, the earth completes one counterclockwise revolution about its north-south axis, whereas the plane of oscillation remains constant once the pendulum has been set in motion. In 1851, the French physicist Léon Foucault (1819–1868) actually carried out this experiment with a pendulum 67 meters long and a bob weighing 28 kilos. Understandably, he chose for his experiment the Parisian Panthéon rather than the North Pole. Geometrical reasons associated with this move meant that the ostensible rotation of the plane of oscillation slowed down,[63] but even so, the result was unequivocal and impressive. For the observer in the Panthéon, the plane of oscillation did indeed appear to rotate, and this effect could only be accounted for if you accepted as a fact that the earth rotates about its own axis.

Answering one question all too often raises a host of others. So, if the earth revolves, what does it revolve around? What kind of standpoint would allow us to see the earth rotating about its own axis? Where would we need to be to see it traveling through space at ten times the speed of a cannon ball in its trajectory around the sun? Where would we have to take up our post to observe the solar system hurling through space at 250 kilometers per second in the direction of the constellation of Cygnus? There is—this much at least becomes clear from Foucault's pendulum experiment—no justification at all for the belief that it might be possible to chart the universe with geometrical means from a terrestrial basis. The solid ground for which Archimedes was searching that would have enabled him to move the earth at will is forgotten as we face ever-retreating horizons.

84 Rainer Maria Rilke: "The heavens, vast and full of awesome reticence, / brimming with space, an overwhelming, stored-up world...."

85 The principle of Foucault's pendulum.

Overpowered by the dimensions of the cosmos that seem to poke fun at any kind of understanding, we need once again to ask ourselves what status human beings can seriously claim. Surely the survivor of a shipwreck has an infinitely better chance of being spotted in the vast expanse of the Pacific than poor Earth has with all its creatures in the hollow depths of the cosmic void. How can cries of newborn babies, groans of anguish or howls of pure ecstasy have any significance outside their immediate environment? On the almost completely empty stage of the universe, how can the spectacle of human history be anything more than an insignificant sideshow beside the cosmic drama of supernovae and black holes?

86 A detail of our galaxy: every scintillating dot is a star comparable to our own sun.

Earth's dimensions are so trivial when set against the vast distances between the stars or the sheer immensity of a galaxy that our very existence can appear to be devalued. Does it not take a reckless, even pathologically exaggerated sense of our own importance for us to persist—in view of those cold, inhospitable and endless spaces—in assigning meaning to our lives, value to our will and purpose to our actions? What hope is there of attracting the attention of the *ranks of angels* and of persuading them to give ear to what our feeble voices are saying?

It may very well be that a defeatist attitude like the one suggested results in resignation and, at best, in indifferent detachment vis-à-vis our place in the cosmos. There are, of course, the frugal among us who take pride in making do with little and delight in having at least calculated the extent of our isolation in a universe that seems quite literally to be Godforsaken. Another and outstandingly sympathetic project to reinfuse existence with meaning was proposed by the French writer Albert Camus (1913–1960). In his *The Myth of Sisyphus,* he outlined the need for all of us to take up a heroic stance and to rebel against the absurdity of existence.[64]

We seem to have boxed ourselves into the tightest of corners. Let us now see whether we can escape by a rigorous application of our theoretical tools: geometry and the nexus between space and numbers. If we can thereby prove that there is nothing inherently dreadful about space as such, it would be absurd to give way to depression simply because space seems so immense.

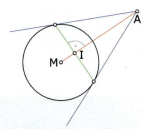

Let us start with a time-hallowed limbering-up exercise: drawing a circle. In the present context, our circle represents a cross section of the earth through its center. Through an arbitrarily chosen point outside the circle, which we will call *external point*, we draw two lines tangent to the circle and a third one connecting our external point with the center of the circle. Then we connect the two tangential points with a straight line. The point of intersection

87 Reflection in the circle: every external point (A) corresponds to an internal point (I) and vice versa.

between this line and the line connecting the external point and the center of the circle, which mirrors the external point inside the circle, we will call *internal point*. It is obvious that there is an unambiguously defined *internal point* to correspond to any external point we might happen to choose. The reverse construction[65] will again transform any internal point into its twin outside the circle. Almost any point inside the circle is a potential candidate to represent its external twin point, the one exception being the exact center of the circle. This method of construction, which geometricians have called the *reflection in the circle*, is seen to establish an unambiguous, two-way relationship between all points outside a circle and all those inside it other than its exact center.

This reflection technique can also be applied to a sphere, which in our context will model the earth. Let there be a random point outside the globe, an *external point*, and let that point be connected with the earth's center by a level plane. The plane's intersection with the globe will yield the points required for the construction just outlined; again, the result will be an internal point that corresponds unambiguously to its external twin. The position of the intersecting plane, incidentally, is apparently of no consequence provided it fulfills the requirement that it establish a connection between the external point and the center of the circle.

Such a reflection in the sphere results in a *hollow world*. In this world, we would be living inside a concave earth rather than on the crust of a convex one. The whole universe of interest to astronomers and stargazers—sun, moon, planets, stars, constellations, galaxies and the rest—would be encapsulated inside the earth's cavities, whereas what is to us the earth's interior would extend outward.

At first sight, the hollow-world theory has little to commend it. However, in order not to weaken its case unduly, we must consider that all of the laws of physics would also be subject to the process of *reflection in the*

88 A photo of the earth as seen from the moon taken from Apollo 8. It seems to prove that the earth is a sphere floating in the universe whose surface is the stage for the drama of human existence…

89 … yet the same picture would make sense in terms of a hollow-world theory. The sun, depicted here as located near the *center of the earth* only illuminates half the inside of the sphere as the rays of light travel along circular paths through the center. Most of the light rays emanating from the moon are lost in the night of the *center of the earth*, leaving only a few actually to reach the earth's surface.

circle. Bodies would appear smaller the farther they moved away from the concave surface of the earth toward the center of the globe; on reaching that center, they would shrink to zero. This also applies to distances; all distances ever calculated by astronomers can be accommodated within the sphere. Light rays would not be traveling in a straight line. A ray connecting two internal points would travel along a circular line sweeping toward the central point of the sphere. Its speed would diminish as it approached the center, and it would disappear completely on reaching it. This explains why standard experiences would remain unchanged in a hollow world:

> An observer at the top of a very tall tower would still only perceive part of the earth's surface within a circular horizon.

> A satellite high up in the sky would still be transmitting photos suggesting the earth was a sphere.

> The sun, which in a hollow-world universe circles the earth in 24 hours, would alternately bathe each of the earth's hemispheres in daylight as the sunlight travels along circular paths. One hemisphere would be lit when the other was shrouded in darkness.

To sum up, a scenario can be contrived in which all of the laws of physics are as valid in a hollow-world universe as they are in our more familiar one, in which sunlight travels along straight paths at an unchanging speed.[66]

There is one crucial implication, though, that needs to be spelled out. The universe, which astronomers present to us in terms of an incomprehensibly vast space, is reduced in the hollow-world theory to touchingly humble proportions, whereas the earth looms very large indeed. So what is the universe like in reality? Is it limitless or contained, compacted or open?

It goes without saying that we are not advancing a plea in favor of the hollow-world theory. There is not a single reason why we should seriously consider adopting it. We have become so used to the idea of existing *on* a globe and of an immense universe stretching away from the earth's surface that we are not going to accept being stood on our heads by the hollow-world theory. The conclusion we want to draw from this discussion is on a different level altogether. The hollow-world theory is highly speculative but *so is the mainstream universe with its gigantic dimensions*. In the last resort, it too is a theory, the imaginative product of human intellectual endeavor. The crucial point is this: there is not one piece of solid evidence—nor can there ever be such a thing—to prove to us whether the dimensions of the universe are *really and truly* the ones measured by astronomers. In just the same way, there is no solid evidence that the *real* cosmos is experienced when the astronomer's universe is subjected to the *reflection in the circle*.[67] So we have no way at all of making a rational choice between an immeasurably vast universe and the domesticated one seen in the *reflection in the circle*.

In the last resort what we want to say is this: the space that is amenable to being charted with the help of Bhaskara's figure is not a *given* reality, it is an intellectual construct. *Space* is a category to help us put the chaos of our optical perceptions in order. It is numbers, as units of length measurement, as unit squares or unit cubes, that enable us to put this category systematically and methodically to work.

Nowhere is this principle realized with such clarity as in the works of the French mathematician, scientist and philosopher René Descartes (1596–1650). Descartes' discovery of the system of *coordinates* made Bhaskara's figure available for an endless range of surveying tasks. We may summarize it as follows:

> A point in space is chosen completely at random to serve as the origin of the system of coordinates.
>
> Three straight lines at right angles to one another are then drawn passing through that point. These are the three axes of the system of coordinates.
>
> At equal distances from the origin, unit points are marked off on each of the three axes, with each point signifying *one* on its axis. To make

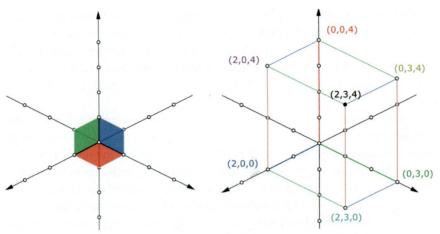

90 The three axes of a Cartesian system of coordinates.

91 The point (2, 3, 4) in the Cartesian system of coordinates.

a distinction of the visual representations of *one* possible in abstract calculations, Descartes proposed rendering the unit point on the first axis as (1, 0, 0), on the second as (0, 1, 0) and on the third as (0, 0, 1).

Each of the three axes of the system of coordinates is a scale with its origin as the visual representation of *zero* and the unit point as that of *one*. It can accommodate any meaningful multiple of *one*; *three*, for example, refers to the point along an axis of the system of coordinates that lies three times as far from the origin as the unit point.

When Descartes refers to a point such as (2, 3, 4), the way to locate what he had in mind is as follows:

Two steps from the origin along the first axis get us to the point (2, 0, 0).

Three steps in a direction parallel to the second axis, and we arrive at the point (2, 3, 0).

Finally, four steps in a direction parallel to the third axis will get us to the point defined above as (2, 3, 4).

Systematically adding step to step along the axes and their parallels will open up three-dimensional space in its entirety.

How the universe could be surveyed using the Cartesian (a word taken from *Cartesius*, the Latin form of Descartes' name) system of coordinates can be demonstrated by adapting the method that the ancient astronomer Aristarchus of Samos used for calculating the distance separating earth and sun:

Let the center of the moon be the origin of the system of coordinates and let the first axis extend from the center of the moon to our eye.

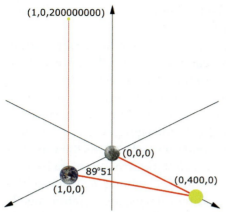

92 Earth as the point (1, 0, 0), the moon as the point (0, 0, 0) and—exactly at half moon—the sun as the point (0, 400, 0). The whole universe has been captured in numbers, even the star in its position vertical to the plane of the earth, the moon and the sun, which is an *astronomical* distance away.

Let the point of contact with our eye be the first unit point. The unit of length measurement is therefore the distance from the observer's eye to the center of the moon.

Precisely at half moon, the sun's rays hitting the moon (the origin of our system of coordinates) form a right angle with this axis of the coordinates. The line connecting the center of the moon with the sun can therefore serve as the second axis, and the third one follows from the first two with logical necessity. Measurement of the angle at which our eye perceives the sun relative to the first axis will indicate, on the basis of Bhaskara's figure, the sun-moon distance in terms of the number of units—where one unit equals the earth-moon distance.

Aristarchus' inaccurately measured 87° led him to the erroneous result of 19 units. A more accurate measurement would have given him 89°51', which would have improved the result considerably to 400 units. In Cartesian diction, the coordinates of the three celestial bodies are as follows:

the moon—or, more precisely, the center of the moon—(0, 0, 0),

the earth—or, more precisely, the observer's eye—(1, 0, 0)

and the sun either (0, 19, 0) in Aristarchus' version or, using modern measurements, (0, 400, 0).

Yet no matter whether we consider Aristarchus' version (0, 19, 0) or the updated (0, 400, 0), the sense of the vastness of space has evaporated. Similarly, a star overhead in a position vertical to the triangle formed by sun, moon and earth might have the coordinates (1, 0, 200,000,000). Even in such a case what has become of the awe-inspiring depths of the universe that exercised such fascination on Giordano Bruno? They have been reduced to mere disconnected numbers, with little of the numinous or the sublime about them. Certainly, as Galileo knew and Giordano did not, they are not worth going to the stake for.

Similarly, the question why the universe has three dimensions rather than four or any other number is not really germane. It does not address the problem that the questioner has in mind. Faced with a task such as drawing more than three straight lines through one point, any one of which forms a right angle with any of the others, we would probably conclude that it cannot be done. Yet this conclusion is not a guarantee that it cannot be done. *Mathematically* speaking, no *a priori* obstacle blocks the way. Nor will it do simply to say there can be no such thing. We cannot embark on a serious investigation by simply stating that some things *are* whereas others *are not*. The question concerning the three dimensions therefore merely addresses the limits of our understanding, not the nature of space as such. The actual number of dimensions remains a completely moot point.[68]

As a mathematician, Descartes has bequeathed us many outstanding achievements. As a philosopher, he has any number of brilliantly formulated ideas to his credit. However, his view of the cosmos as a whole can at best be called unsatisfactory. In a domain that he had made uniquely his own by applying mathematics to it, he remained fettered to the tradition of Giordano by his persistence in regarding space as a *given* reality *out there*.

Descartes believed that the world consists of two categories of objects. One category, *res extensa*, comprises *extended objects*, which are under the sway of thrust, pressure and all the other laws of physics. Such objects are defined by the space that they fill and the space that surrounds them. By *res cogitans*, his second category, Descartes understood the *thinking* subject, which differs from *res extensa* in that it does not extend in space. Its defining characteristic is the *soul*, which is independent of space. Angels and human souls belong to the category of *res cogitans*. Inanimate objects such as stones and so on are obviously *res extensa*, but so are the *soulless* plants and animals. For Descartes, plants and even animals function on the most basic, purely mechanical level of thrust and pressure. The callousness of Descartes' convictions comes into full view when he listens to the cry of pain of a tortured animal with utter indifference, as if what he was hearing was no more significant than the ringing of an alarm clock. He fails to detect any traces of *consciousness*, let alone of *soul*, in the *mechanism* of a suffering animal.[69]

To be able to introduce the distinction between *res extensa* and *res cogitans*, Descartes unquestioningly posited space as existing independently of our thought processes. Space and thought, we might say with only a touch of caricature, form a pair of opposites for Descartes. In truth, as we have seen from our discussion, this is by no means the case. Space is not the *opposite* of thought, it is its *object*. If the world needs to be split into pairs of opposites, as Descartes suggested, *res extensa* versus *res cogitans* is less satisfactory as a pairing than *res cogitata*, the category of *thought about* objects, versus *res cogitans*, the *thinking* subject. Space clearly belongs within *res cogitata* and may be expressed by means of numbers.

The myth of the almost unbearable vastness and emptiness of space — that desolate scene of all our aimless wanderings — turns out to be a phantasmagoria, just as much an illusion as the blue color of little Giordano's distant mountains. The enormous cosmic distances ultimately owe their existence only to our ability to calculate them. Outside numbers, there is nothing in space for us to discover; within mathematical thought, there is room for all the vastness of the universe.

Leibniz

Numbers and Logic

No thinker was ever more taken with the power of logic and no other scholar was more universally erudite than Gottfried Wilhelm Leibniz (1646–1716). Equally at home in philosophy, history, theology, linguistics, biology, geology, mathematics, logic, law and diplomacy, he had set his heart on discovering a universal method for attaining reliable knowledge and understanding the nature of the cosmos. Thomas Hobbes' (1588–1679) flash of insight that all thinking is calculating made a deep impression on

93 Gottfried Wilhelm Leibniz.

94 Thomas Hobbes.

the young Leibniz, who went on to explore the possibility of a *lingua characteristica universalis*, a universal language informed by symbolic logic, which would make logical errors every bit as obvious as arithmetical errors.

It is deeply ironical that one so hopeful as Leibniz was to advance the cause of mathematics and add to its glories should have landed it, through that very optimism, in a state of unparalleled confusion. Numbers are at the heart of Leibniz' project to demonstrate that logical thought and mathematics share the same essence. They are equally central to the discovery 300 years later of the limits of calculation, which has posthumously created huge question marks against Leibniz' optimism.

To find out more, let us start by considering just why we use the notation for numbers that we learned in childhood.

Long familiarity with the conventions of our numerical notation system has blinded us to the ingenuity of the solution devised by Arab mathematicians—or rather their Indian teachers. We will appreciate the elegance of the Arabic place value system more readily when we compare our current numerical notation to the much more cumbersome systems adopted by other civilizations. The Romans, for instance, used selected letters for their notation system. As we all know, I = 1, V = 5, X = 10, L = 50, C = 100, D = 500 and M = 1000. Numbers not expressible through a single cipher are notated as combinations, and the symbols are repeated as needed. The ciphers are written in decreasing order from left to right,[70] so that, for instance, DCCC = 800, MDCCLXXXVIIII = 1,789, MDCCCCXIIII = 1,914. The letters

IIII, XXXX and CCCC were often replaced by IV, XL and CD; VIIII, LXXXX and DCCCC by IX, XC and CM. Yet even if 1,789 and 1,914 are spruced up in this way (to MDCCLXXXIX and MCMXIV), they are still not easy to read.

Because we are accustomed to the Arabic system, this method of numerical notation makes calculation laborious. Additions such as XVII + XXXIII, XXXIII + LXVI, subtractions such as L – XXXIII, XCIX – LXVI are not at all easy if we attempt to do the calculations without first converting them into our system. Multiplication in Roman numerals is a real chore,[71] and division

95 DCCC.

96 MDCCLXXXVIIII.   **97** MDCCCCXIIII.

is so complicated that students in the Middle Ages had to wait until they entered the university to be initiated into its mysteries.

The Indian-Arabic notation system in use today has its own numerical symbols, or *digits*, for the first ten numbers (assuming that you start counting from zero), the digits 0, 1, 2, 3, 4, 5, 6, 7, 8, 9. These digits, shifted to the left as required to incorporate the powers of ten (1, 10, 100, 1000, 10,000, …) are used as building blocks to express any given number. Here are two examples:

$$2{,}309 = (2 \cdot 1000) + (3 \cdot 100) + (0 \cdot 10) + (9 \cdot 1)$$

or

$$65{,}537 = (6 \cdot 10{,}000) + (5 \cdot 1{,}000) + (5 \cdot 100) + (3 \cdot 10) + (7 \cdot 1).$$

Faced with the task of explaining our *base ten* system to an ancient Roman, we could proceed as follows:

1. We must agree that the digits 1 to 9 are alternative versions of the first nine Roman numerals so that 1 = I, 2 = II, 3 = III, 4 = IV, 5 = V, 6 = VI, 7 = VII, 8 = VIII and 9 = IX.

2. We also will have 0 represent *nihil* (nothing). (We do not know anything else about the Indian mathematician who was behind the introduction of the zero some 1,500 years ago but what a stroke of genius it represents! The whole sophistication of the place value system turns on it.)

3. For the number following 9—(9 + 1)—we will write 10, with the digit 1 moved one place to the left (into what we will call the *tens* column) representing one times ten and 0 in the base position (the *units* column) representing *nothing*.

Next, we might explain to our Roman colleague how to convert a Roman numeral such as MMMDCIX into our system, which we may now designate *decadal*, or *base ten*.

We go to the right-hand end of the Roman number, provided it repre-sents a number smaller than ten. In our example, IX is converted into the digit 9, which goes into the units column, giving us 9.

1. We must subtract IX from the Roman numeral, leaving MMMDC.

2. Next, the Roman number undergoes a metamorphosis: every M sym-bol is replaced by a C, every D by an L, every C by an X, every L by a V and every X by an I. The new shape of MMMDC is therefore CCCLX. This *metamorphosis* is, of course, a division by ten.

3. We must then treat CCCLX according to Step 1, chiseling away at the right-hand end looking for a singles digit between zero and nine. Because there is no such digit in our random example, we will put a zero into the tens column of our base ten number, which now reads 09.

4. Then, we divide CCCLX by ten and, following the procedure in Step 2 above, we arrive at XXXVI.

5. Chiseling away again at the right-hand end, we will obtain 6 for the hundreds column. The base ten number is now 609.

6. Metamorphosing XXX to III, we obtain 3 for our thousands column and are left with 3609.

$$\text{MMMDCIX} = 3609$$

For added clarity, the process is summarized in the following table:

M → C		I = 1 II = 2 III = 3 IV =4
C → X	D → L	V = 5 VI = 6 VII = 7
X → I	L → V	VIII = 8 IX = 9 and 0
MMMDC → CCLX		MMMDCIX = MMMDC + 9
CCCLX → XXXVI		CCCLX = CCCLX + 0
XXX → III		XXXVI = XXX + 6
		III = 3

We have devoted so much space to a simple conversion because a simi-lar procedure is being used today in a place value system that is unfamil-iar to most people—though it is commonplace in the world of information technology (IT). It seems certain that our ten fingers led to our adoption of the base ten system, but a differently aggregated base would probably serve our purposes equally well. Leibniz first proposed the introduction of a place value system with the smallest possible number of different digits, which we now call the *binary* system. This system actually uses only two digits: 0 for zero and 1 for one. *Two*, the number following *one*, is notated in binary as 10; this binary combination replaces the base ten digit 2. The digit 1 goes into the twos column, and the zero in the units column again indicates *nothing*. These digits, shifted to the left as required to incorporate the powers of two (1, 2, 4, 8, 16, 32,…), are used as building blocks to express any given number.

To obtain the binary equivalent of any number, we proceed much as we did when we showed our Roman colleague how to dispose of MMMD-CIX. There is one difference though: the distinction between odd and even numbers now plays an important role in the context of a division by two. We will take 75 as our example:

1. We start by noting that 75 is an odd number. We therefore select the Leibniz cipher 1 to fill the units digit of our new binary number: 1.
2. We subtract 1 from 75 and are left with 74. This number can be divided by two and thereby metamorphoses to 37, again an odd number. To fill the twos column, we select the Leibniz cipher 1: 11.
3. We subtract 1 from 37; 36 is then divided by 2 and metamorphoses to 18. As 18 is an even number, we select 0 for the fours column: 011.
4. We divide 18 by 2 and obtain 9, an odd number, which yields the Leibniz cipher 1 for the eights column: 1011.
5. We subtract 1 from 9 and divide the result, 8, by 2 to get 4. As 4 is an even number, 0 goes into the sixteens column: 01011.
6. This 4 metamorphoses to another even number, 2, so we obtain another 0 for the thirty-twos column: 001011.
7. Finally, the last metamorphosis produces the number 1, which means 1 goes into the sixty-fours column: 1001011.

$$75 \text{ base ten} = 1001011 \text{ binary}$$

Once again, the process is summarized in the following table:

Halve	$0 = 0$ $1 = 1$
$74 \rightarrow 37$	$75 = 74 + 1$
$36 \rightarrow 18$	$37 = 36 + 1$
$18 \rightarrow 9$	$18 = 18 + 0$
$8 \rightarrow 4$	$9 = \ 8 + 1$
$4 \rightarrow 2$	$4 = \ 4 + 0$
$2 \rightarrow 1$	$2 = \ 2 + 0$
	$1 = \quad 1$

The reconversion of binary to base ten is even simpler. A binary number, say, 111001 is converted as follows, starting from the units digit:

$$11100 = 1 + (0 \cdot 2) + (0 \cdot 4) + (1 \cdot 8) + (1 \cdot 16) + (1 \cdot 32) = 1 + 8 + 16 + 32 = 57.$$

The binary system is perfectly suited for the inner workings of a computer because 0 and 1 mirror the electromagnetic states of "charged/ uncharged", "polarized/non-polarized" or "conducting/non-conducting". This is why the unit of information in a computer is called a *bit*, which is the abbreviation of *bi*nary digi*t*. It is stored in the smallest available memory space, a space that can only accommo- date either 0 or 1.

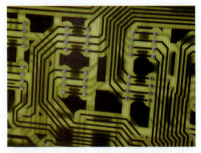

98 Electronic circuits are based on binary logic.

The most obvious drawback of the binary system is that it makes num- bers, as they get bigger, swell disproportionately to unwieldy monsters. The number 3,609, for example, instantly comprehensible in decadal nota- tion and still reasonably transparent as MMMDCIX, runs into a great string of digits in binary: 3,609 = 111000011001. This in itself would be enough to deter anyone from trying to substitute base 2 for base 10 in everyday life. Computers, on the other hand, with their exponentially larger stor- age capacity have no problems digesting yard-long strings of 0s and 1s. A base ten number typed into a computer is instantly converted into binary by an onboard program and is subsequently processed in that form. It is reconverted into base ten for human consumption by the same translation program going into reverse. Needless to say, the user remains unaware of all these comings and goings.

This drawback is far outweighed by the reduction of all binary addi- tions, subtractions, multiplications and divisions to a few very basic rules. All we need to know to perform any addition is this:

$$0 + 0 = 0, \quad 0 + 1 = 1, \quad 1 + 0 = 1, \quad 1 + 1 = 10.$$

To add two numbers—say, 75 = 1001011 and 57 = 111001, the two num- bers are written one below the other with their place values in vertical alignment. Starting with the units column, we calculate the sums of the digits in each column. In the case of 1 + 1, we write down the 0 and "carry" 1 to the next column. This will yield the result

$$\begin{array}{r} 1001011 \\ +\ \ 111001 \\ \hline 1000100 \end{array}$$

and 1000100 = (128 + 4) = 132, which is indeed the correct sum of 75 + 57.

It is clear, moreover, that standing these same rules on their head will enable us to carry out subtractions of smaller numbers from bigger ones:

$$10 - 1 = 1$$
$$1 - 0 = 1$$
$$1 - 1 = 0$$
$$0 - 0 = 0.$$

To subtract $57 = 111001$ from $75 = 1001011$, the two numbers are written one below the other with their place values in vertical alignment. Starting with the units column, we calculate the differences of the digits in each column. If the calculation concerned is $0 - 1$, we replace the 0 of the number at the top by 10 (we "borrow" 1 from the next column). As part of the next step, we "carry" 1 and add it to the lower number. This will yield the result

$$
\begin{array}{r}
1001011 \\
- \ 111001 \\
\hline
10010
\end{array}
$$

and $10010 = 16 + 2 = 18$ is indeed the difference between 75 and 57.

Incredible though it may seem, multiplication in the binary system is an even simpler process than addition or subtraction, or, as we might say, "one-times-one" is even simpler than "one-plus-one" or "one-minus-one". The four rules look like this:

$$0 \cdot 0 = 0, \quad 0 \cdot 1 = 0, \quad 1 \cdot 0 = 0, \quad 1 \cdot 1 = 1.$$

For multiplications with 10, 100, 1000, 10,000, ..., we simply add one, two, three, four, ... zeroes to the number being multiplied. To multiply our two numbers $75 = 1001011$ and $57 = 111001$, all that is required is to form the sum of the following numbers (because 75 in binary notation = 1000000 + 1000 + 10 + 1):[72]

$$57 \cdot 75$$

$$
\begin{array}{r}
= \ 111001 \cdot 1001011 \\
111001 \\
111001 \\
111001 \\
111001 \\
\hline
1000010110011
\end{array}
$$

The addition yields the binary behemoth of 1000010110011, which is indeed $75 \cdot 57 = 4{,}275$ because $1000010110011 = 1 + 2 + 16 + 32 + 128 + 4{,}096 = 4{,}275.$[73]

	0000	0001	0010	0011	0100	0101	0110	0111	1000	1001	1010	1011	1100	1101	1110	1111
000.	NUL	SOH	STX	ETX	EOT	ENQ	ACK	BEL	BS	TAB	LF	VT	FF	CR	SO	SI
001.	DLE	DC1	DC2	DC3	DC4	NAK	SYN	ETB	CAN	EM	SUB	ESC	FS	GS	RS	US
010.	SP	!	"	#	$	%	&	`	(	)	*	+	,	-	.	/
011.	0	1	2	3	4	5	6	7	8	9	:	;	<	=	>	?
100.	@	A	B	C	D	E	F	G	H	I	J	K	L	M	N	O
101.	P	Q	R	S	T	U	V	W	X	Y	Z	[	\	]	^	_
110.	`	a	b	c	d	e	f	g	h	i	j	k	l	m	n	o
111.	p	q	r	s	t	u	v	w	x	y	z	{	|	}	~	DEL

99 The ASCII-code: the column on the left gives the three highest binary digits of the code; the line at the top supplies the next four binary digits.

The binary system of Gottfried Wilhelm Leibniz is able to accommodate all numbers (starting from zero), the simple rules required for adding, subtracting, multiplying and dividing and a great deal besides. Combining batches of eight bits into packages that we call *bytes* makes it possible to encode any text in binary form; every character on the typewriter keyboard is assigned an unambiguous, reversible seven-digit binary code number. Taking into account accents, punctuation marks, additional characters and non-printable control characters, we will find 128 code numbers perfectly adequate for the purpose, from 0 = 0000000 to 127 = 1111111. An example is the code known as *ASCII*.[74] In this code, the letters *d*, *n* and *a* are assigned the numbers 100 = 1100100, 110 = 1101110 and 97 = 1100001.[75]

A "check bit" is attached to each of these seven-digit binary numbers, its purpose being to ensure that the seven bits of the encoded character are made up to an eight-bit byte that contains an odd number of ciphers of 1.[76]

It follows that 1100001 0 1101110 0 1100100 0—the binary number 110000101101110011001000 = 12,770,504—is equivalent to the word *and* for the ASCII-trained computer.

Any text—regardless of whether it is trite copy for a soap ad or a Paul Celan poem of daunting complexity, whether it is the three-letter word *and* or a book with the epic dimensions of the Bible—is reduced to a string of binary code by the computer.

Of course, converting text into binary code only scratches the surface of the actual potential of computers. Indeed, so complex are the tasks within their capacity that it is sometimes difficult not to give in to the illusion that these machines are capable of actual autonomous *thought*.

Leibniz was the first to realize that there was more to the binary system than the possibilities it offered for encoding. It was also highly suitable for the delineation of problems in logic. Although he failed to pursue this *logical calculus* in detail—another almost 150 years passed before George Boole (1815–1864), the English logician and mathematician, picked up that thread—he did play around with the idea of

100 George Boole.

making the binary digits 0 and 1 represent *true* and *false*. If we develop this idea of Leibniz', we could equip every logical statement with an evaluative bit:[77] *0* if the statement is to be marked false; *1* if true.

Let us consider as a simple example the two statements "Romeo loves Juliet" and "The Montagues hate the Capulets", both of which, going by Shakespeare's *Most Excellent and Lamentable Tragedie of Romeo and Juliet*, are factually true. From a standpoint of mere logic, this is, of course, not necessarily so. Someone not familiar with Shakespeare's play has no way of telling.

So, let us say that both statements could be either true or false. Let us then examine the statement "Romeo loves Juliet *and* the Montagues hate the Capulets". We may calculate the truth content—or lack thereof—of this *combined* statement by means of the evaluative bits tagged on to the component statements. All that is required is a multiplication of the evaluative bits:

False	True	False	True	
0	1	0	1	Romeo loves Juliet.
$0 \cdot 0$ $= 0$	$1 \cdot 0$ $= 0$	$0 \cdot 1$ $= 0$	$1 \cdot 1$ $= 1$	Romeo loves Juliet and the Montagues hate the Capulets.
0	0	1	1	The Montagues hate the Capulets.

If both component statements are held to be false (expressed through the evaluative bits 0), multiplying these evaluative bits gives the calculation $0 \cdot 0 = 0$, which shows that the compound sentence organized around the coordinating conjunction *and* is also false.

If one of the component sentences is true and the other false: the calculations $0 \cdot 1$ and $1 \cdot 0$ both result in 0, the tag for false. The combined statement is therefore false.

Only if both of the statements are true do we arrive at the calculation $1 \cdot 1 = 1$ and a situation where the combined statement is true.

As it is, we know from Shakespeare that both statements "Romeo loves Juliet" and "The Montagues hate the Capulets" are true and must therefore get the tag *1*. Given the logical force of *and*, the compound sentence "Romeo loves Juliet and the Montagues hate the Capulets" is also true. The multiplication $1 \cdot 1 = 1$ reflects precisely that state of affairs.

101 Romeo loves Juliet ...

In short, not only does the binary system convert the word *and* physically into a string of 0s and 1s, but it also even unleashes its logical potential and makes it available for the purposes of calculation—it calls for a multiplication of evaluative bits. This is only one example of the logical force of conjunctions being made visible in calculation, precisely in the way envisaged by Hobbes.[78]

The function of logic, as far as Leibniz was concerned, can eventu-

102 ... and the Montagues hate the Capulets.

ally be advanced by the construction of a universal calculation program:

> Type in an argument.
>
> The argument is automatically converted into binary code.
>
> After a great deal of number-crunching in an endless series of interlocking steps, the evaluation 1 or 0 will let you know whether the conclusion is true or false.

In his speculations, Leibniz anticipated artificial intelligence by several centuries. It is only thanks to the development of computers capable of processing millions of bits per second that we are able to put his prophetic vision of logic becoming subservient to calculation to the test. Autopilots in aviation and computer-assisted medical diagnostic methods are just two examples of our increasing reliance on computer programs to help us make the right decisions in a wide range of areas.

In his utopia of a world organized along rational lines, Leibniz wanted to see any type of conflict—legal disputes as well as political ones—settled not in an arena of clashing interests but through the objective evaluation of arguments cast in the form of numbers. "Let's sit down together and work out the result" was to be a motto as applicable to conflict resolution as to mathematics itself. The idea is still a little fanciful today, but it is nowhere near as utopian as it was in Leibniz' day. Perhaps the last thing with which people in the Baroque era wanted to see their vibrant world twinned was numbers; today, on the other hand, we have come to believe, perhaps a little foolishly, that practically anything at all can be made to yield to some form of quantification.

Nowadays, the symbol of music is the CD, which acts almost as a deep freeze, preserving even the most delicate composition as a gigantic binary number waiting only for the microwave of our music players.[79] Similarly, it seems that any visual or acoustic stimuli can be reduced to digital data, to mere numbers, to monotonous sequences of the same two digits. It is true that these reductions are often riddled with defects and are never wholly faithful reproductions of the various stimuli, but surely no one will deny that today's digital technology presents us with virtual worlds of breath-taking realism. When IT specialists talk about programs or simulations, what they are really discussing—whether they are aware of it or not—are numbers in binary form. Layout programs, digital image processing, the financial services of a bank—it all boils down to inputting strings of binary numbers and to setting a procedure in motion that produces yet another sequence of the digits 0 and 1 as output.

Having said this much, let us pause for a moment. Is it really only a matter of time before Gottfried Wilhelm Leibniz' dream is realized, that dream of discovering a method for solving each and any problem through calculation? Do we have no part to play beyond waiting for a new generation of computers, computers that will process zillions of zeroes and ones in no time at all? Or are there obstacles ahead forever blocking our path to an understanding of the world in terms of formal logic—obstacles that must ultimately banish Leibniz' dream to the historical lumber room reserved for failed utopias?

To understand better what computers are capable of doing, let us now focus on computer programs for a moment. Not all programs are excessively complicated. For instance, the program we outlined earlier on which we can rely to come up with the output of 132 when we input 57 + 75 is an altogether modest affair. Scarcely more complex is the program that delivers the output of 4,275 for the input 57 · 75. However, when we move on to a program for solving the division 57/75, we advance several steps in complexity. This division yields the decimal number 0.76, so the output generated by the program would be the three-digit sequence 0, 7, 6. Another, almost identical program specifies the solution to the reverse division, 75/57. This, however, results in a further complication, because 75/57 = 1.31578947368..., which is an *infinite decimal* number. Accordingly, the output generated by the program consists in a string of numbers that can never end.[80]

To have a uniform format at our disposal to note all kinds of results, we will agree that output always takes the form of an uninterrupted string of digits. Before the program calculates a result, let that string be

$$0, 0, 0, 0, 0, 0, 0, 0, 0, 0, 0, 0, ...,$$

an uninterrupted sequence of zeroes. When the program produces a result consisting of only one number, as is the case with the two simple programs for the addition of 75 and 57 and for the multiplication of these numbers, let that result replace the first of the zeroes:

$$132, 0, 0, 0, 0, 0, 0, 0, 0, 0, 0, 0, ...$$

or

$$4,275, 0, 0, 0, 0, 0, 0, 0, 0, 0, 0, 0, ...$$

When the program produces several numbers as the result, as is the case with the two division programs, *57/75 and 75/57*, many of these zeroes will be replaced from left to right by the numbers obtained. This could result in two (strictly speaking, three) replacements, as in the case of 57/75:

$$0, 7, 6, 0, 0, 0, 0, 0, 0, 0, 0, 0, ...,$$

with the first zero standing as a result rather than a placeholder.

However, there could also be infinitely many replacements, as in the case of 57/75:

$$1, 3, 1, 5, 7, 8, 9, 4, 7, 3, 6, 8, ...$$

Let us give these programs the self-explanatory *names* of 57 + 75, 57 − 75, 57/75 and 75/57. The table below contains the names of these programs and the results in their output line:

Name	Output
57 + 75	132 0 0 0 0 0 0 0 0 0 0 0 0 0 0
57 · 75	4275 0 0 0 0 0 0 0 0 0 0 0 0 0 0
57 / 75	0 7 6 0 0 0 0 0 0 0 0 0 0 0 0
75 / 57	1 3 1 5 7 8 9 4 7 3 6 8 4 2 1

It is crucial for what follows that each program has a specific *number* as well as a *name*, which we call its *program number*.

Recall that for data to be processed by a computer, every input must be converted into binary form. As programs are a form of input, this stricture applies equally to them. The program *57 + 75*, which performs the addition of these two numbers, is stored in the computer's random access memory perhaps under 1001101111 … 001000111, with the three dots representing an apparently haphazard string of zeroes and ones. The onboard conversion program interprets this binary monster as a command to come up with the result *132*. A similar process applies to the programs *57 · 75*, *57/75* and *75/57*.

We realize then that by compiling a list of *binary* numbers,

$$0, 1, 10, 11, 100, 101, 110, 111, 1000, 1001, 1010, 1011, \ldots,$$

we are also compiling a list of all conceivable *program numbers* (admittedly including many that will be unmanageable for the conversion program because not all binary numbers double as meaningful program numbers— a problem the computer solves by simply generating a result consisting of zeroes only; this step is classified as a *default*).[81]

Let us assume that our four programs, *57 + 75, 57 · 75, 57/75* and *75/57*, are assigned the (arbitrarily chosen) binary numbers 11 = 3, 101 = 5, 1000 = 8 and 1001 = 9. (In real life, program numbers as short as these would be totally unrealistic. Even the simplest of programs have code numbers with dozens of digits. The wish to keep the apparatus of technical paraphernalia down to manageable proportions also entails spelling out these idealized binary numbers in our familiar base ten notation.) We supplement the previous table with a column listing the program numbers.

Name	Program Number	Output 0.	1.	2.	3.	4.	5.	6.	7.	8.	9.	10.	11.	12.	13.	14.
·	0	·	·	·	·	·	·	·	·	·	·	·	·	·	·	·
·	1	·	·	·	·	·	·	·	·	·	·	·	·	·	·	·
·	2	·	·	·	·	·	·	·	·	·	·	·	·	·	·	·
57 + 75	3	132	0	0	0	0	0	0	0	0	0	0	0	0	0	0
·	4	·	·	·	·	·	·	·	·	·	·	·	·	·	·	·
57 · 75	5	4275	0	0	0	0	0	0	0	0	0	0	0	0	0	0
·	6	·	·	·	·	·	·	·	·	·	·	·	·	·	·	·
·	7	·	·	·	·	·	·	·	·	·	·	·	·	·	·	·
57 / 75	8	0	7	6	0	0	0	0	0	0	0	0	0	0	0	0
75 / 57	9	1	3	1	5	7	8	9	4	7	3	6	8	4	2	1
·	10	·	·	·	·	·	·	·	·	·	·	·	·	·	·	·

The lines in the table filled with dots are on standby to receive the programs numbered 0, 1, 2; 4; 6, 7; 10, and the table could be continued 11, 12, …, to accommodate any number of conceivable programs.

One more preliminary note: it is common practice among IT mathematicians to start counting with zero. Therefore, from here on we will refer not only to the *zeroeth* program (the program with the program number *0*) but also to the *zeroeth* number in the output line (the number at the beginning of the line). For instance, in the preceding table, the zeroeth number in the output line of program number 5 is the number 4,275. Similarly, the fourth number in the output line of program number 9 is 7 (rather than 5, as we would probably have assumed until we recalled that we also have to start counting from zero in the output line).

Now we come to the salient point that we have been working toward. We propose to fill one of the vacant program slots with a program called *Eubulides*.[82] The name has been chosen in homage of the man who provided, in an age that antedated computers by some 2,300 years, the intellectual spark for such a program to be created. This is what Eubulides is all about:

> Its zeroeth output number is the zeroeth output number of the program with the program number 0 augmented by 1.
>
> Its first output number is the first output number of the program with the program number 1 augmented by 1.
>
> Its second output number is the second output number of the program with the program number 2 augmented by 1.

This continues in the same manner—for instance, its tenth output number is the tenth output number of the program with the program number 10 augmented by 1.

Expressed in more general terms, here is what Eubulides does:

> It scans the numbers 0, 1, 2, 3, 4, …
>
> In doing so, it adds 1 to the zeroeth, first, second, third and fourth output numbers of the program it is scanning.
>
> It then puts these numbers into its own output line as the zeroeth, first, second, third, fourth, … numbers.

Admittedly, Eubulides looks bizarre at first sight. What is it purpose? As far as programming is concerned, it is almost as undemanding as our multiplication program. Like any other program, Eubulides has to be assigned a program number. As before, assigning it a realistic program number is irrelevant to our present intentions; for simplicity's sake, we will assume that its number is 11. Our table has now been modified to include Eubulides, complete with program number and output line:

Name	Program Number	Output 0.	1.	2.	3.	4.	5.	6.	7.	8.	9.	10.	11.	12.	13.	14.
.	0	.	.	.	.	.	.	.	.	.	.	.	.	.	.	.
.	1	.	.	.	.	.	.	.	.	.	.	.	.	.	.	.
.	2	.	.	.	.	.	.	.	.	.	.	.	.	.	.	.
57 + 75	3	132	0	0	0	0	0	0	0	0	0	0	0	0	0	0
.	4	.	.	.	.	.	.	.	.	.	.	.	.	.	.	.
57 · 75	5	4275	0	0	0	0	0	0	0	0	0	0	0	0	0	0
.	6	.	.	.	.	.	.	.	.	.	.	.	.	.	.	.
.	7	.	.	.	.	.	.	.	.	.	.	.	.	.	.	.
57 / 75	8	0	7	6	0	0	0	0	0	0	0	0	0	0	0	0
75 / 57	9	1	3	1	5	7	8	9	4	7	3	6	8	4	2	1
.	10	.	.	.	.	.	.	.	.	.	.	.	.	.	.	.
Eubulides	11	.	.	.	1	.	1	.	.	1	4	.	■	.	.	.

Each slot in the output line of Eubulides corresponds directly to the other programs in our table. Where a program is on standby, the corresponding slot in Eubulides contains a dot. After the numbers of these programs are known, they are as readily processed as the numbers 1, 1, 1 and 4 that Eubulides has processed, all of which represent the relevant output value (printed in red in our table) augmented by 1.

The crucial problem is indicated by the red square in slot 11 of Eubulides' output line. How is the result for slot 11 to be obtained? Eubulides checks slot 11 against its program number 11 and adds 1. Obviously, program number 11 is Eubulides itself so the command for it is to alter its own eleventh output value—which is self-contradictory nonsense and locks Eubulides in a deadly clinch with itself. Clearly, such a state of affairs must not be allowed to occur. Yet how is it to be avoided when Eubulides is formulated in such utterly unequivocal terms?

The Austrian mathematician Kurt Gödel (1906–1978) and his English contemporary Alan Turing (1912–1954) found an answer whose presentation calls for a few preparatory remarks. If we return to our table, programs with program numbers lower than Eubulides, such as program number 10, will behave as follows:

> One millisecond, one tenth of a second and one second after the start of the program, the computer comes up with the zeroeth, first and second numbers of the result: 3, 1, 4.
>
> The third number, 1, appears 10 seconds later.
>
> One minute elapses before the arrival of 5 to fill slot four.
>
> We wait another ten minutes before the arrival of output value 9 for slot five.

Filling slots six, seven and eight with **2**, **5** and **5** takes an hour, ten hours and two days, respectively.

We have to wait five days before the computer deigns to come up with **3** as the output value for slot nine.

Then the computer starts to calculate the tenth output number of the program with the program number ten. It counts and counts, churning around its numbers for days, weeks, months, years.

We wonder whether it will ever stop calculating the tenth output number of the program with the program number ten. Maybe there is a *loop* hidden in this program that will cause the computer to go on shunting its numbers around in a circle without ever completing its calculations. If this is the case in the present context, there will be no more meaningful results, and the output line will freeze with

$$3, 1, 4, 1, 6, 9, 2, 6, 5, 3, 0, 0, 0, 0, 0, 0, \ldots$$

Yet it is equally possible that there is no loop in the program bringing about a *default* in its calculation of the number to fill slot ten. In such a case, all we have to do is wait—after 12 or 12,000 or 12 billion years of ceaseless calculations, the computer will finally produce the result **5** to go into slot ten.

Name	Program Number	Output														
		0.	1.	2.	3.	4.	5.	6.	7.	8.	9.	10.	11.	12.	13.	14.
·	0	·	·	·	·	·	·	·	·	·	·	·	·	·	·	·
·	1	·	·	·	·	·	·	·	·	·	·	·	·	·	·	·
·	2	·	·	·	·	·	·	·	·	·	·	·	·	·	·	·
57 + 75	3	132	0	0	0	0	0	0	0	0	0	0	0	0	0	0
·	4	·	·	·	·	·	·	·	·	·	·	·	·	·	·	·
57 · 75	5	4275	0	0	0	0	0	0	0	0	0	0	0	0	0	0
·	6	·	·	·	·	·	·	·	·	·	·	·	·	·	·	·
·	7	·	·	·	·	·	·	·	·	·	·	·	·	·	·	·
57 / 75	8	0	7	6	0	0	0	0	0	0	0	0	0	0	0	0
75 / 57	9	1	3	1	5	7	8	9	4	7	3	6	8	4	2	1
xyz	10	3	1	4	1	5	9	2	6	5	3	?				
Eubulides	11	·	·	·	1	·	1	·	·	1	4	?				

Eubulides, as Gödel and Turing discovered, can function only when it is equipped with a *loop recognition program*. Such a program is designed to check other programs and indicate via 0 or 1 whether the program can generate a result or whether it is going to get itself trapped in an endless loop.

Equipping Eubulides with such a program would enable us to see at a glance whether it was worth waiting for the tenth output number of program number 10: the loop recognition program would warn us of the existence of an unending loop by adding 1 to the default 0. Eubulides could then continue its calculations successfully until it reaches its own program number, which would result in the impasse just described.

In the formal logic of thinking in binary terms, however, there is no such thing as an impasse, which makes us realize there can be no loop recognition program. This insight has been given the names of *Gödel's Incompleteness Theorem* or *Turing's Negative Solution of the Entscheidungsproblem*. It states that no program can

103 Kurt Gödel.

attain to the universality needed to indicate for every program whether it will achieve a result on a foreseeable time scale or whether it will continue forever in an endless loop.

What does this mean to us? Simply, that no matter how sophisticated computers may become in the future, this particular problem will always prove one too many for them. This delivers us from any quasi-religious belief in the omnipotence and omniscience of computers.

One obvious objection to this conclusion may be that the arguments which purport to prove the fallibility of computers are all based on the bizarre Eubulides program, which is, in any case, utterly irrelevant to the role computers play in real-life situations. The problem may well be insoluble, but it has nothing to do with the applications for which we need computers in logic.

The reason why it is necessary to rebut even such an apparently trivial objection is this: Eubulides is by no means as bizarre or irrelevant as a casual observer might think. In fact, if we formulate the definition of Eubulides in more general terms, it will be seen to be of paramount logical significance. Eubulides' most distinctive characteristic is its purpose of effecting a change in the results of all conceivable programs. If Eubulides succeeds in its self-destructive mission, our dilemma will become acute as the program attempts to effect a change in its own output; it would then be in open contradiction with itself.

We realize we are on logical terrain here that bears a strong resemblance to that posed by the two statements that we examined earlier: "Romeo loves Juliet" and "The Montagues hate the Capulets". The driving force behind Shakespeare's drama is, of course, that Juliet *is* a Capulet and Romeo *is* a Montague so we end up with a straight contradiction; the analogy with Eubulides is clear for all to see. The statement "The Montagues hate the Capulets" is both true *and* false, because Romeo loves Juliet. The truth content of the first statement is altered by the second. In the formal logic of zeroes and ones, the statement "The Montagues hate the Capulets" is capable only of one inference concerning Romeo: Romeo cannot love Juliet. His love of Juliet is a logical impossibility. And yet Romeo *does* love Juliet. He is contradiction incarnate, precisely what makes zero-one logic wilt.

After all, it is not only drama that is fueled by contradiction. Contradiction is the very stuff of which life, as opposed to logical dreams, is made. Without it, life, deprived of all its challenging unpredictability, would quite simply not be worth living.

The *logical identity axiom* (which asserts an object's identity with itself) is so close to tautology that it appears unassailable. As far back as the sixth century BC, however, the Greek Pre-Socratic philosopher Heraclitus begged to differ, when he declared it was impossible to step into the same river twice. What he meant, no doubt, was that both the river and the bather become different entities; the river water is constantly being renewed, and the bather ages into a different person. Experience sides with Heraclitus. No form of being is so immutable as the strict logic of thinking in ones and zeroes would have us believe.

Similarly, the principle of the excluded third, which postulates that a logical statement is either true or false and that a third possibility can be ruled out, is unlikely to provoke vehement opposition. However, there is good reason to call that principle into question, too. Is it true that every computer program will either come up with an answer or lose itself in an endless loop? If nothing else makes us pause, at least *Gödel's Incompleteness Theorem* should make us wary of answering rashly in the affirmative.

It is truly amazing that Leibniz' philosophical-mathematical mind should have given birth both to inexorable binary logic and to a branch of mathematics diametrically opposed in its essence: the branch we call *calculus*. Calculus is uniquely capable of capturing change, movement and flux—in a word, the world of Heraclitus.

To demonstrate that Leibniz broke with the principles of strict logical consistency in what was arguably his greatest achievement, let us consider two points situated on a curve. These two points may be either identical or different. On the basis of the axiom of the excluded third, the statement *The two points are identical* is either true (meriting the evaluative bit 1) or false (0).

If the statement is false and the two points are different, they provide an unambiguous definition for one particular straight line: the line that passes through both of them. This straight line obviously intersects the curve at the two points. Interpreting it as a tangent to the curve is surely out of the question as a tangent touches a curve at one point and one point only.[83]

However, if the statement is true and the points are in fact identical, then this self-identical point certainly does not provide a definition for only one straight line. On the contrary, there will be an infinite number of straight lines that pass through it. Which one of these innumerable lines should be called the tangent to this point on the curve?

Leibniz provided an answer to this question by first considering the two points as different and then switching to treating them as identical at what he considered an appropriate juncture in his argument. To justify this dereliction of logical duty, he resorted to the dubious explanation that the two points must be thought of as *only just* different, differing only by *evanescent increments*.

A brilliant philosopher like Leibniz' Irish contemporary George Berkeley (1685–1753) had little time for such an untenable position and assailed it with mordant irony: "And what are these evanescent increments? They are neither finite quantities, nor quantities infinitely small, nor yet nothing. May we not call them ghosts of departed quantities?"

While it is obvious that Leibniz did not play by the rules and

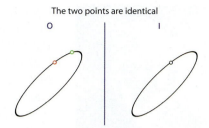

104 Two points on a graph can be identical to or different from each other.

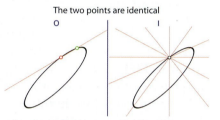

105 If the two points are identical, a random number of straight lines can be drawn through the spot marked by them. If the two points are different from each other, only one straight line passes through them, which is not a tangent to the curve.

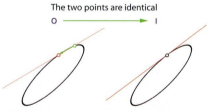

106 Leibniz starts out by considering the two points as different from each other in order to be able to draw an unambiguously defined straight line through them and then switches to considering them as coinciding in a single point in order to be able to treat this straight line as a tangent to the curve.

107 Luitzen Egbertus Jan Brouwer.

Berkeley's criticism is justified, Leibniz' calculus has proved to be one of the most outstanding intellectual achievements of all time. The logical anomalies that Leibniz introduced into his argument—in full awareness of what he was doing and following a finely tuned set of rules—proved all-important stepping-stones in the development of a descriptive method whose dynamics match that of the vibrant world it seeks to model. It is no exaggeration to suggest that our modern age was uniquely shaped by calculus in technological, social and cultural terms.

If we take the eminent Dutch mathematician Luitzen Egbertus Jan Brouwer (1881–1966) as our guide, we will not be apologetic about denying pole position to logic. Brouwer argued that logic must always be subordinated to our direct intuitive knowledge of numbers. It is on this, not logic, that calculus is ultimately grounded. We simply have to jettison such constricting logical principles as the principle of the excluded third if we do not want to restrict our access to the rich harvest offered by the limitless field that is numbers.

Laplace

Numbers and Politics

In nationwide election campaigns, televised debates involving candidates, journalists and presenters are important milestones. It is incredible how many numbers are bandied about by the participants on those occasions. In one 90-minute confrontation some time ago, as many as 196 numbers were pulled out of various hats in efforts to substantiate the speaker's political program, attack opponents and win favor with potential voters. Think of it, 196 numbers in 90 minutes—more than one number for

108 Pierre Simon Laplace.

every 30 seconds of discussion. This did not include those vital groups of numbers that we use to refer to years, which are so important in political debates. Here are the nearly 200 numbers that were cited:

```
50000000000 40 10 2 25000000000 43 1000  100000000 10000000000
43,5  11  35000000000  15000000000  3,5  8  22000  4 77000000000 6
8000  5  200000000000  8  1200  50  40  4  1500  90000 5000000000 5
20000  180000000  4900000  8500   8700  34231000000  12000000000
20000000000 12000 22051000000 41,7  120000 95 300 43 6600 10000
74000000000 90000 11000000000 500 3  2200 40000000000 11000 70
25000000 17000000000 20 30 11500 5600000000 40 60 8400 55 50 40
3  65  22000  6  10000  39 8000  102000000000  3  60  50000000000
122000000000  5  75  12  142000000000  ¼  85 48  1600000000000 ½
11000  50  50000000000  ¾ 12000 10000000000 5 2 2 5000000000 9 3
40000000000 4,5  1 2 5000000000 1,8 2 1/3 200000000000 15 2 7 100
4 8 8 1000000000 15 122 2500000000 9 20 140 5000000000 25000000
200  120  5  50000000  10  10000000000 3000000000  7000000000 15
12000000000 1600000000000  3  3600 15000000000 120000000000
190000000000  20  700000000000  125000000000  50000000000  21
10000000000  15000000000 7000000000  15 15000000000  0,5  28
9000000000 1000 2,5 1300 1000000000 240000 3 65000 1000 35000 8
11000000000  2  2500000000  60  30000000000  3  1500000000  9 980
10000  4,7  100  200  5  2 58  10000  1000  15000000000  23  1  55
60000000000 19 2 22700000000 10 18 3 7
```

Ranging from the monstrous to the puny, 1,600,000,000,000 to ¼, these numbers were tossed around pell-mell in the discussion in the order shown in the preceding figure. Of course, not even the most attentive audience could possibly take all this in. Nor can it be believed that sharing information, the purpose these numbers are claimed to serve, is the uppermost consideration in the contestants' minds. It is more likely that their true purpose is obfuscation, because slinging around numbers dotted with millions and billions has little to do with sharing information—it is rather like peacocks spreading their tails in their mating rituals. Nor is it very helpful when participants in the discussion feel called upon to patronize their audience by such ploys as: "So if our national debt of 1.6 trillion were a stack of 1,000 dollar bills, it would rise to a dozen times the height of Mount Everest."

It comes as no surprise that the politician who benefited most from the debate in our example made the cleverest and most sparing use of numbers, and he was the one who went on to win the election. In the debate, he referred to numbers only four times, and he did so very early on. The budget, according to him, had a $50 billion gap, which he proposed to plug in three installments of $25, $10 and $15 billion, obtained from three different sources. The simple addition of 25 + 10 + 15 was the only arithmetical challenge offered to the viewers. Beyond that, the future victor refrained from all references to numbers whatsoever, preferring instead to taunt his opponents[84] with, "Our figures are on the table, where are yours?"

Numbers are a source of awe, fascination and dread for us. The successful campaigner is someone who is able to exploit to his or her advantage our mixed feelings about numbers, feelings that seem to comprise equal parts of attraction and revulsion. How is it that most people's feelings about numbers are so ambivalent? Probably the chief reason is that numbers create an illusion of establishing indisputable certainty and absolute accuracy (which we crave) but they do so by cold-heartedly producing a solution that is never more than quantitative (which we resent). Counting itself has no limits. The bigger numbers get, the more they transcend our imaginative faculties. At the same time, the actual borderline between what we can and cannot visualize remains elusive.

Pierre Simon, Marquis de Laplace (1749–1827), one of France's most brilliant mathematicians, physicists and astronomers, has left us two fundamental ideas that throw our ambivalent attitude toward numbers into sharp relief: the idea of the daemon named after him and the idea of probability.

Laplace's daemon endowed numbers with divine attributes. To understand this, we need to remember Newton and his laws of celestial mechanics: Newton used the laws of mechanics to account not only for the movements of the planets around the sun and of the moon around the earth but also for the trajectories of comets and other celestial phenomena. Newton formulated his laws to explain even the very minor deviations that the various celestial bodies show from the elliptical orbits originally assigned to the planets by Kepler. He was aware that gravity was as much at work between these bodies as it was between each of them and the sun. This deviation was particularly in evidence in the elliptical orbits of Jupiter and Saturn, the two biggest planets. Ultimately, it seemed to point to the eventual destabilization of the whole solar system. Two complete revolutions of Saturn around the sun correspond to five of Jupiter's. This means that the two planets come into relatively close proximity on a regular basis, increasing the eccentricities of their orbits every time.

109 Sir Isaac Newton.

110 Jupiter and Saturn appear to threaten the stability of the planetary system.

Newton, believing that this threat to the stability of the solar system required divine attention every 50,000 years or so, tried to get a theological perspective on an otherwise calamitous situation in a letter to his pupil Samuel Clarke (1675–1729): "It is no lowering of God, but the true glorification of his works if one says that nothing proceeds without his eternal management and supervision."

Newton's somewhat helpless notion of God being constantly at the ready to prevent his planets from making a nuisance of themselves excited Leibniz' ridicule:[85]

> Sir Isaac Newton and his followers have also a very odd opinion concerning God's creation. According to their doctrine, God Almighty has to wind up his watch from time to time: otherwise it would cease to move. He had not, it seems, sufficient foresight to make it a perpetual motion. Nay, the machine of God's making is so imperfect that he is obliged to clean it now and then in an extraordinary service, and even to mend it, as a clockmaker mends his work; who must be so much more unskillful a workman, as he is oftener obliged to mend his work and set it right.

A century later, Laplace substituted a mathematical explanation for Newton's theological one to account for the stability of the solar system. The orbits of Jupiter and Saturn were discovered to be close to, but not identical with, a simple integer ratio of 5:2, which means that the point of their closest approximation shifts over centuries. Applying his "perturbation technique", Laplace was able to demonstrate that the two planets reverted to their original orbits after approximately 900 years. God had lost his role as protector and preserver of the solar system. During a lecture on celestial mechanics, Laplace was asked about God's place in his universe by Napoleon and famously answered: "Sire, I have no need of that hypothesis."

In the eyes of Laplace, the universe, from the smallest grain of dust to its most powerful suns, was one soulless automaton:

> We may regard the present state of the universe as the effect of its past and the cause of its future. Conceive of an intellect which at a certain moment knew all forces that set nature in motion, and all positions of all items of which nature is composed: if this intellect were also vast enough to submit these data to analysis, it would embrace in a single formula the movements of the greatest bodies of the universe and those of the tiniest atom; for such an intellect nothing would be uncertain and the future just like the past would be present before its eyes.

In this quotation, in which Laplace introduces us to his *daemon*, he boldly replaces Almighty God with Almighty Numbers. All that is required for the daemon to know is six numbers about each and every atom (in the

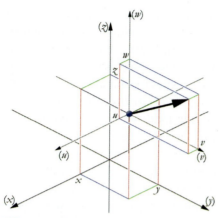

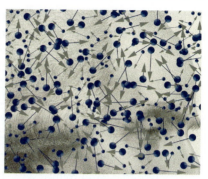

112 The monstrous welter of data required for the calculation of the universe according to Newton's laws is too much for our minds and for our computers—but not, according to Laplace, for his fictional *daemon*.

111 Knowledge of the location and the velocity of each particle is the basis from which to predict the future.

Laplacean sense of elementary points, which were thought to be the constituent elements of all objects in the universe):

> the three spatial coordinates that indicate how far *in front, to the right* and *above* the atom is relative to a fixed point of reference, and

> the three coordinates of its velocity that specify how fast the atom moves *forward, to the right* and *upward*.

On the basis of Newton's laws, these numbers would enable the daemon to calculate the state of the cosmos at any given time in the future or in the past.

Laplace was aware that it was impossible for human beings ever to have all the numbers needed for these operations at their disposal—there are simply too many atoms for that. Even if these data were available and an array of tens of thousands of today's most powerful supercomputers were used, the computational time for the project would span several human lifetimes. This is why Laplace hit on the idea of the daemon. What is beyond the reach of us mere mortals may yet be possible. His fictional daemon was a warranty in Laplace's eyes that the cosmos functions, at least *in principle*, no differently from a fully comprehensible machine.

Progress in physics brought about the end of Laplace's daemon.[86] The decisive moment came when the German physicist and philosopher Werner Heisenberg (1901–1976) formulated his *Uncertainty Principle*. After this, it was no longer possible to think in terms of atoms as *points* whose spatial and velocity coordi-

113 Werner Heisenberg.

nates could be determined at any given moment in time with a predefined degree of accuracy. We may summarize as follows: Newton's laws, which Laplace used as his frame of reference, are applicable only to *non-existent* objects. Only in highly simplistic circumstances—such as the large clumps of bonded atoms with which we deal in celestial mechanics—is Newton's mathematical system reliable. For all other applications, it has been superseded by the far more complicated *quantum* mathematics, which reduces Laplace's daemon to a decrepit chimera.

Even for someone who believed that quantum theory would not be the last word on the matter and preferred to go on thinking of atoms in the Laplacean manner, Laplace's daemon would still be up against essentially insurmountable difficulties. After all, it simply will not do to have only approximate specifications (giving an accuracy of, say, three decimal places) of the space and velocity coordinates of all the atoms in the universe. It is in the nature of calculus (developed jointly by Newton and his adversary Leibniz) that, in many of its formulae, the most diminutive change in the numerical material will have devastating consequences for the final result. Even if just one of the spatial coordinates of just one atom were measured to be 3.141592653589793 rather than 3.141592653589792— a difference of 1 in the fifteenth digit after the decimal point—the result would be a different cosmos.

114 A butterfly flapping its wings *out of sync* ...

115 ... may trigger a tornado in another part of the world months later.

The *butterfly effect*, as it is sometimes called, says that a butterfly flapping its wings in a Brazilian rain forest may cause a tornado in Florida months later. This is the degree of instability mirrored in the differential equations that model weather changes. A mathematical theory, developed and intensely popular in the last decades of the twentieth century and called *chaos theory*, is particularly efficient at visualizing cloud formations, coastlines, the growth pattern of tree-like organisms and many other natural and social phenomena. It depends largely on how hypersensitive equations are to small changes in initial conditions.

It is quite obvious that for Laplace's daemon to discharge its more than Herculean labor with any real success, spatial coordinates and velocity vectors must be specified with a higher degree of accuracy than 15 decimal places. Yet even 15 million or 15 quadrillion accurate decimal places in this all-engulfing flood of numbers would by no means be sufficient for the computation of the universe. Let us examine the preconditions for the daemon to become operational:

> A mind-boggling number of data sets—six times as many as there are atoms in the universe.

> Each measurement in each of these sets would have to be *absolutely accurate*.

But is *absolute* accuracy possible? If so, how is the daemon to achieve it? For a start, it must be placed in a position where it chooses at its discretion to how many decimal places each coordinate is calculated. Whatever choice it makes, it still cannot be equal to its task. The daemon can never be sure that its computation of the world will accord with reality as long as it is based on arbitrarily truncated numerical material. It is always possible that a more exacting definition of accuracy will lead to a radically different result. The daemon must therefore demand that not just arbitrarily many but *all* of the infinitely many decimal places be put at its disposal, a demand that is simply impossible to fulfill.[87] As we have already pointed out, nowhere do we actually encounter the infinite, and the very notion of a workable realization of the infinite is patently absurd.

As we have seen, there can be no such thing as Laplace's daemon, not even in principle, not even as an intellectual construct. We are therefore mercifully spared the dreadful consequences that would follow from its existence, namely that

> free will is illusory,

> the distinction between good and evil is immaterial,

> there is no such thing as moral responsibility for one's thoughts, words and actions, and

> we lack the moral right to base our children's education on ideals or to punish evil-doers for their crimes.

Yet at the same time, we would be wrong to assume that it follows from the daemon's non-existence that numbers are not, after all, *almighty*. That idea is now transferred to a different setting, which was moreover also provided by Laplace.

Only Laplace's daemon would be capable of predicting the value of the roll as a die is being

116 Dice as the symbol of "chance".

cast. In its absence, that value is determined by chance. Chance may be blind, but that does not place it beyond the reach of calculation. Laplace realized that the symmetry of a die—none of its faces differs from any of the others (except in its markings[88])—means that each of the six possible results is equally likely, and each has the same probability of occurring. The probability of throwing a particular value, be it 1, 2, 3, 4, 5 or 6, is therefore $1/6 \approx 0.1667 \approx 16.67\%$. What is meant by the term *probability*?

Laplace might have given an answer[89] along the following lines:

> Let us assume that a casino pays out \$3,600 to anyone who predicts the outcome of a die roll correctly.
>
> As long as the stake required is higher than \$3,600/6 = \$600, the casino will turn in a profit in the long run (i.e., over a protracted series of bets).
>
> If the stake required is lower than \$600, the casino will eventually go bust.

The game typically associated with casinos is not dice, but roulette. However, and despite its much more sophisticated styling, roulette does not actually differ much from dice. Here is a summary:

> The gamblers place their bets.
>
> The croupier calls out, *"Rien ne va plus"* ("No more stakes will be accepted"), and throws an ivory ball into a slowly spinning, cauldron-shaped roulette wheel.
>
> After a phase of circling the rim of the wheel freely against its direction of rotation, the ball bounces to and fro like mad and comes to rest at last in one of the pockets numbered from 0 to 36 nestling against the rim.

The probability that the ball will land in a particular pocket, such as the 14 preferred by James Bond, is therefore $1/37 \approx 0.027027$. In percentage terms, this is equal to a little over 2.7027%. This value should be seen in the same light as the 1/6 probability of a certain number of spots ending on top in the throw of a die. So let us assume that a casino pays out \$3,600 to a gambler who has predicted exactly which pocket the ball would land in before the croupier called, *"Rien ne va plus."* As long as the stake required is higher than \$97.30 ($\approx$ \$3,600/37), the casino will turn in a profit over a protracted series of bets. Casinos that pay out \$3,600 on a smaller stake than this will eventually go bust.

In fact, casinos allow themselves a safety margin and require a stake of \$100 before they will pay out \$3,600 on a successfully predicted pocket. The bank kept by a casino following these guidelines is sure to make a profit in the end.

117 Roulette.

Bets are available in roulette other than the all-or-nothing (in roulette parlance, *en plein*) bet just examined. Eighteen of the pockets from 1 to 36 are red (*rouge* in roulette parlance),[90] while the other eighteen are black (*noir*). Zero is green. The probability of the ball ending up in a red pocket is therefore 18/37 ≈ 0.4865 ≈ 48.65%. To put it another way, there are 18 chances that the ball will come to rest in a red pocket and 19 that it will not. So, 18 of the 37 possible outcomes can be described as favorable to the gambler who places his bet *rouge*.[91] If the ball lands in *rouge*, all players who have bet on that color win double their stakes. In the long run, the casino makes a healthy profit because in 10 million games, the ball may be expected to land in *rouge* very much fewer than five million times. If we multiply the various figures, we will see that the probability in mathematical terms (correct to seven significant figures) is that *rouge* will come up in only 4,864,865 of these 10,000,000 games.

Most roulette wheels in the USA have a second zero pocket marked 00. For wheels such as these, the arithmetic is as follows:

Chance of success for an *en plein* bet: 1/38 ≈ 0.26316 ≈ 2.6316%.

Stake at which the casino breaks even on a $3,600 payout: $94.74 (≈ $3,600/38).

Chance of success for a *rouge* bet: 18/38 ≈ 0.4737 ≈ 47.37%.

Possible outcomes: 38, of which 18 are *favorable*.

Probable *rouge* in 10,000,000 games: 4,736,842.

But what on earth does gambling have in common with the mechanics of the skies or even with politics? More than one might think, when we take into account the non-existence of the Laplacean daemon.

The attempt to calculate exactly all the movements of all the celestial bodies in the solar system, beginning with the sun and its eight planets, passing on to demoted Pluto, the moons of the planets and the asteroids and ending with the comets, is a project doomed to failure from the outset. Yet even if we were to succeed in these calculations, we would then have to take into account the influences on our solar system of other stellar systems and other galaxies throughout the universe.

Although the immediate effects of these influences are very small indeed, they may well be cumulative, creating trouble for us in the long term. In other words, even though the solar system appears to function like the works of a gigantic clock, the Sisyphean labor of calculating its movements by computation can never be accomplished, no matter how powerful are the computers enlisted for the task. What can be accomplished, though, are highly probable predictions regarding the future and an equally reliable historical analysis of the past. Such predictions have a much higher probability coefficient than, say, the weather forecast, which can claim reliability only for a week or so in advance. However, like the weather forecast, they belong to the order of probabilistic statements. We can go to bed every night assured that the sun will rise again in the morning, but it would be nearer the truth if we were to refer to the next sunrise, not as a certainty, but as a probability of perhaps 99.99999999997%.[92]

We have accepted that weather forecasts hinge on probability. However, accepting that the safety devices of large-scale industrial plants, such as nuclear power plants, guarantee their smooth functioning with a

118 Sunrise is all very well, but can it be taken for granted?

119 The failsafe functioning of technological installations can only be guaranteed with a high degree of probability.

probability that ultimately falls short of 100%, even if only by a very small margin, is much more worrisome.

Similar concerns can be raised regarding medicine. Every case of dialysis, a medical procedure needed to remove waste such as urea from the blood because of kidney failure, has a so-called *mortality coefficient*. This alarming phrase specifies the probability of the outcome that a patient will die as a result of the medical intervention. Of course, no effort is spared to keep that mortality coefficient as low as possible, but it cannot be eliminated entirely. At present, this amounts to 0.001%, which means the degree of probability of a patient dying during dialysis is 800 times higher than the probability of someone winning the Euromillions lottery after buying only one ticket. To win that game, you need to match five numbers from 1 to 50 plus two *power stars* drawn from a group of nine.

Of course, calculations of probability do not affect the fate that is of personal concern to an individual—in our example, the patient worrying about his dialysis. For the individual whose life hangs on the thread of his doctors' medical skill, the probabilistic statement has no relevance; at best it plays the role of a placebo and has a tranquilizing effect on the nervous patient before his medical ordeal. Yet you cannot be 0.001% dead—either you survive the procedure, as do 999,990 out of a million patients, or you die, in which case, you are 100% dead. For the same reason, it is useless to point out the extremely unfavorable odds to people about to sit down at a roulette table or to purchase lottery tickets. Their hope that they belong to the lucky few or that they will be the one and only as in fairy stories or Hollywood films means that they will not want to listen.

The mortality coefficient has the last word on the fate of patients only in a limited, *political* sense. The hospital administration keeps records of all cases of mortality during dialysis treatment. If the numbers rise significantly above the percentage specified by the mortality coefficient, stringent quality checks of clinical wards are instituted that will lead to identifying and eliminating trouble spots. In the same *political* sense, life insurers calculate the premiums for policyholders. Insurance agents always try to create the impression that their company has the personal welfare of each individual client at heart. The real picture, however, is governed by the large number of the insured, which relegates the individual to the rank of a statistical cipher among many others. Insurance companies weight their premiums by calculations of probability, to which the very large numbers involved lend an astonishing degree of precision.

If these hard facts give you a headache, pharmaceutical companies have the answer. Every drug has its own coefficient of success—the probability that it will do what it says on the box. Let us assume a drug company submits two of its headache drugs to trials: Alpha and Beta. In the case of Alpha, 192 out of 242 women participating in the trial register an improvement in their condition; for Beta, the figures are 357 out of 510 women. The probability of the drug alleviating the women's headache is therefore

for Alpha, 192/242 = 80%; for Beta, 357/510 = 70%.

It is obvious that Alpha is more effective among women than Beta. Now the company submits the two drugs to trials among groups of male sufferers. In the case of Alpha, 288 out of 720 men in the trial feel beneficial effects; for Beta, the relevant figures are 51 out of 170 participants. The probability of the drug alleviating the men's headache is therefore

for Alpha, 288/720 = 40%; for Beta, 51/170 = 30%.

It is true that men are less ready to respond favorably to either drug, but the truth remains that Alpha is more effective than Beta for them also.

Is Alpha therefore really the better drug? This is far from certain. Even the data presented here are far from trustworthy. Let us consider another perspective. Alpha was submitted to trials involving 240 women and 720 men (a total of 960 persons), and 192 + 288 = 480 felt relief due to the drug. The trials for Beta involved 510 women + 170 men = 680 persons altogether; 357 of the women and 51 of the men felt a beneficial effect, which is a total of 408 persons altogether. If we conflate the results from the women with those from the men, the probability that the prescription of these drugs will be successful is as follows:

for Alpha, 480/960 = 50%; for Beta, 408/680 = 60%.

Suddenly, Beta emerges as the more promising candidate—and no one, literally no one, can presume to make a rational decision, in spite of the data presented here, on which is the more effective drug.[93]

	Alpha		Beta	
Women	192/240	80%	357/510	70%
Men	288/720	40%	51/170	30%
Sum total	480/960	50%	408/680	60%

120 A paradoxical result: when results are considered separately under gender headings, Alpha seems to be the more efficacious drug; overall and for the population as a whole, Beta seems to be more promising.

The upshot of such situations where the real significance of the numbers involved remains doubtful is frequently that statistically correct data are hijacked and used for partisan purposes. Supporters of Alpha will point with satisfaction to the gender-sensitive calculation of Alpha's probable efficacy, whereas the Beta camp will highlight the overall figure. Far from being used to determine true merit, they are instead served up garnished with rhetorical decor to dress up as irrefutable fact what is desirable from a partisan point of view.

Let us remember that most statements involving numbers that we encounter in the political and other arenas come from the field of probability. It is more than questionable whether the people who habitually spike their arguments with numbers acknowledge that they have any responsibility for such data, still less for the way they are assessed. Among the most predictable examples are high-profile reports in the media on the efficacy of some experimental cancer drug, in which statistics are regularly invoked, fostering vain hopes among patients and their families.

This is not the place to discuss in detail the shabby little tricks used to load statistics in support of some predefined goal. There have been cases in which increases in the turnover of a company of 1.00% in one year and of a further paltry 1.31% in the ensuing year have been sold to unsuspecting shareholders by a cutthroat management as a growth of 31%. Similarly, a homeopathic healer may claim that his or her herbal remedies are more successful than hi-tech medicine. What he or she fails to mention is that the critically ill patients are routinely deposited in traditional intensive-care wards before they become a liability to his or her statistics. Again, a city administration starved for success stories may keep the number of drug-related deaths artificially low by placing overdosed patients on life-support machines until they die of heart failure, which can then be passed off as a *natural* death. For those interested in the subject, there is a fast growing number of well-written and amusing books on "how to lie with statistics".[94]

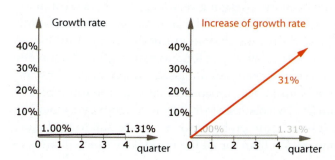

121 What is in reality a barely perceptible growth rate is sold as an impressive increase.

We ought to touch briefly on one important concept in statistics: *data correlation*, which is frequently used to imply that there is a causal link between two observed phenomena. A classic example of a correct positive correlation is the spread of probabilities for developing cancer of the lungs. Comparing non-smokers and smokers, you realize that the chances of succumbing to lung cancer are significantly higher for smokers than for non-smokers. Statistics, by bearing out this positive correlation between smoking and lung cancer, underscore the intuitive hypothesis that the constant irritation of the lungs due to inhaled smoke must contribute to the development of cancer.

However, there are numerous positive correlations that involve no causal link whatever. Anecdotal evidence suggests that whenever particularly large numbers of storks appear in those regions of Austria where the birds have their habitat, there is a corresponding upward blip in human birth rates. Should this lead us to conclude that babies are indeed brought by the stork? On a more serious note, this type of reasoning shows its unacceptable face when populist politicians exploit the apparent positive correlation between the percentage of migrants in an urban population and the crime rate for their xenophobic policy proposals. Big cities are equally attractive to migrants and to criminals—even if all migrants were perfectly law-abiding, this particular positive correlation would remain.

Therefore, we should be well advised to distrust all claims in which researchers, usually in such fields as sociology, education or psychology, can offer nothing better than positive correlations between the phenomena they have "studied" as evidence for their alleged discovery of causal links. The recognition of a positive correlation has no explanatory power whatever. Its existence is at best an indication that the phenomena in question may indeed be linked in ways that deserve careful study. Politicians who have nothing more persuasive than correlations to support their arguments are quite simply not engaging with reality. We can only hope that their target audience is perceptive enough not to be impressed by their shadow-boxing.

In calculations of probabilities, juggling very large numbers is common. Unfortunately, large numbers tend to send our minds into spasms, so this juggling can often produce results that seem utterly counterintuitive. A good example of this is the probability of ten successive spins of a roulette ball landing in ten different pockets. An unbiased person might think this was fairly high. If we were to ask someone to jot down ten numbers at random ranging from 0 to 36, chances are that these ten numbers would all be different. In fact, it is far more probable for a roulette ball to land more than once in the same pocket than for all the pockets in which it lands to be different.[95] This answer[96] proves surprising for most people, because they are simply not used to dealing with numbers as very large as those required in this particular calculation.[97]

What is even more worrisome is that it is often difficult to translate bald figures into language that reflects their significance. A simple yet highly pertinent example will illustrate this point.

What, you might ask, is the average monthly take-home pay (income after taxes) in any given country? To find as accurate an answer as possible, you would first obtain the relevant data from the country's Bureau of Statistics. These data would typically take the form of individuals per specified income bracket: the number of people earning between $0 and $499, those earning between $500 and $999, between $1,000 and $1,499 and so on. Next, you might put these data onto a bar chart or *histogram*. The height of the individual bars represents the number of the country's employees who earn each specified monthly income; the relevant income is noted in Figure 122 below the lines demarcating each income bracket.

One objection that might be raised immediately concerns the arbitrary demarcation of income brackets: $0 to $499, $500 to $999, $1,000 to $1,499, which introduces a powerful bias from the start. The bar representing the average income bracket of $1,250 includes employees who make little more than $1,000 and those who make just below $1,500—a considerable difference—which this histogram ignores.

However, for this example, this objection will be overruled. The distortion caused by the bracketing together of comparable income classes could be lessened through the introduction of a larger number of narrower income brackets represented by a larger number of bars on the histogram; these could be used to model reality more accurately. Instead, we want to address a much more sensitive issue, to which a histogram based on narrower income brackets is practically irrelevant. In fact, it is easier to address this issue by means of a histogram based on a simplified distribution of income.

Let us consider a fictional country in which the number of salary-earners rises in proportion to income in the bracket $0 to $999 but falls in proportion to income in the bracket $1,000 to $5,000. What is the average wage in this fictional country?

The first answer you are tempted to give would focus on the income earned by the largest group of employees. As is shown by the income distribution graph in Figure 123, the income in question is $1,000.

Yet here an objection needs to be made. Surely a better measure of the average income is that sum which has half the number of employees earning more than or equal to it and the other half earning less than or equal to

Number of income earners

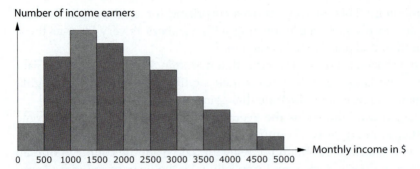

122 Distribution of the income for employees in a fictional state: the height of the bars indicates the number of individuals per income bracket; incomes are notated below the lines separating the brackets.

Number of income earners

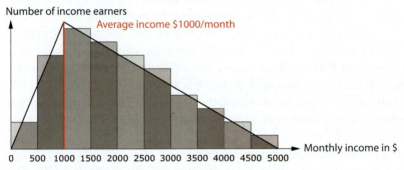

123 In this simplified chart of income distribution, most employees have an income of $1000.

it. Looking at this type of average income graphically, we proceed by taking the total area below the income distribution line and divide it in two.[98] This geometric calculation yields a substantially higher average income: $1,837.32, shown in Figure 124.

There is, however, a third kind of average income. You multiply each income bracket by the number of salary-earners in that bracket, add up all these results and divide the total by the total number of salary-earners. What we are effectively doing here is pooling the total sum earned and redistributing it to all salary-earners in equal portions; each such portion represents an average income. Statisticians call this the *arithmetic mean* of all monthly incomes—in our fictional country, it amounts to exactly $2,000. (See Figure 125.)

Which of these three amounts—$1,000, $1,837.72 or $2,000, all obtained in such sensible ways and yet all so different—represents the *true* average income? Is there a single, exclusive answer to that question? In practice, which of the three definitions is preferred depends on the political agenda

124 According to the same simplified chart of income distribution, one-half of the employed population earn less than $1837.72; the other half earn more.

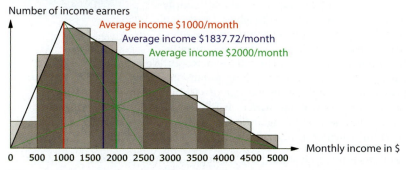

125 According to the same simplified chart of income distribution—and if every employee were to earn the same income—everyone would be receiving $2000.

followed by the person posing the question, and whether he or she intends to follow this agenda in a populist or a responsible manner. This holds true not only for fictional countries but also for all societies and not only for average incomes but for all politically relevant statistical data.

We may conclude that numbers, which all too often bob along helplessly like corks in the daily ebb and flow that is politics, should be considered in two ways. First, they make us aware that beyond all heart-rending individual tragedies, the future of our societies is neither predetermined nor unforeseeable. It is susceptible to intelligent guidance—provided we make a responsible use of numbers. On the other hand, we must bear in mind that any number will remain a sterile husk if it is deprived of the interpretative context in which it can alone unfold its true significance. Forced into an alien context, numbers can lead us in fatally mistaken directions.

Numbers determine our fate. This was Laplace's message. He was right to the extent that politics, in its preoccupation with the majority in a society, ignores the individual. Yet if we accept this statement of his without qualification, mathematics has indeed replaced theology. "What if anything," asks Ludwig Hasler of the respected Swiss weekly, *Die Weltwoche*, "do we stand to gain from this?" He goes on:

> "Why was earth created?" asks Voltaire's Candide. "To drive us mad," comes the answer. Typical Voltaire, one is tempted to say. Yet is the answer therefore necessarily wrong? Because there is always something driving us mad: just now it may be the weather—hurricanes, floods, tornadoes or other disasters, every now and then it may be love, increasingly it seems to be globalization …

126 Voltaire.

> We do have our defences—we do not simply give in to fate and let ourselves be driven mad. We do have numbers: thirteen casualties, five hundred million in damage, twenty-five million cubic feet of windblown wood. We are not saying that such quantification does anything to tame hurricane "Lothar", but it does enable us to cast off the role of victim, and gives us a sense of perspective, control and action. It releases us from confronting raw misery and sets us up again in business as detached professionals, counting, calculating and weighing up assets and liabilities.

> And this is just as it should be. We depend on the belief that we have mastered fate, on our belief in the reliability of technology. When technology lets us down, we have numbers to take care of the fallout. When a Swissair plane crashes with 220 people on board, we are at a loss for words at first but regain our composure when we check the figures: how many deaths in transport-related accidents per billion passenger kilometres? For cars, the figure is 10.6. For planes, 3.1. For trains, 0.6. Nothing to get into a state about. As if anyone ever travels billions of kilometres! A case of numbers putting paid to fear—which is, after all, their speciality. Numbers are unique in being wholly detached from emotions, from debates concerning their meaning, even, in the last analysis from meaning itself. Since nothing is beyond the reach of calculation, numbers have no concept of *chance*.…

> Are numbers then our panacea against going mad? Certainly they are themselves quite capable of driving us mad! Ever since we have set up the economy as the sole dominant force, it has capitalized literally everything—free time, art, emotions—and has us scuttling for our places in the circus of utility, in constant fear that we will count for nothing. Are numbers, our trump card against fickle fortune, about to take on the role of incalculable fate?

> It would appear so. Scholarly talking heads are holding forth about the "dialectic" of numbers. Taking it a step further, they might even refer to numbers as a "metaphysical entity". After all, numbers are already

shouldering God aside to assume special tasks such as the dispensing of payments in compensation for sheer bad luck. Let us remember the latest catastrophes.... First there was the horror, then came the claims. Two million dollars, two hundred million, two billion—no one knows for what purpose, but everyone agrees that this is the way to proceed. "This death is so utterly pointless," complains the father of an Australian canyoning victim. In the meantime his lawyers file a claim for a couple of millions. Numbers as compensation for futility?

Now we understand—and now we don't. What is a life worth—two million dollars or fifty? The question is as meaningless as it is insulting. In the last analysis this is simply not what it is all about. These numbers are seen as compensation not for lives lost but for the blindness of fate, blindness that we find intolerable. The amount, fixed arbitrarily but sufficiently high, is supposed to provide at least symbolic compensation for the arbitrariness of fate. Somehow even a modern pagan feels that there should be some sort of cosmic arbitrator to decree suitable atonement for every sin. We may convince ourselves that God is dead and that metaphysics in its entirety has been consigned to the compost heap of the history of ideas; we are still left with fate, destiny—call it what you will, we feel it needs to be compensated for, understood, and acquiesced in.

So where do we go from here? We do not have contact details for Fate these days. Oedipus in his time was able to see his tragedy as part of a divine drama decreed from above. Job in his downfall was still free to remonstrate with his God. All this is over and done with. As far as our eyes can reach, there is no "higher" system of co-ordinates in sight that would enable us to locate, understand and accept hurricanes and assassinations, mishap and death. The Three Fates of the Ancients have fallen silent and even the God of Christendom, who set us trials here on earth to fit us for heaven, hardly ever returns our calls.

One outcome of all this is that we are less and less ready to accept an accident as simply a chance event. These days there always has to be someone to blame and someone to complain to. If the post of Supreme Judge for Cosmic Justice is vacant, the law courts offer the only hope of redress—redress that takes the form not of consolation but compensation running to increasingly fantastic amounts. This means we have entered the last stage of the evolution of capitalism: the capitalization of fate as orchestrated by trial lawyers. The temple of numbers has been made the venue for the playing-out of compensation claims.

Numbers piled upon numbers. Even ethics is reduced to an exercise in optimizing numbers: the highest good is no more than the promotion of the greatest happiness for the greatest number; the whole business has become nothing more than an exercise in multiplying happiness. Mathematics has become an alternative to theology with numbers, equations and formulae replacing exegesis.

In cosmic theory this is taken for granted. That the parting of light and darkness took place 500,000 years after the Big Bang (something around fifteen billion years ago) may sound like a piece of mythological cosmogony. But there is no myth, no narrative, no heroic saga of a cosmic struggle between good and evil. All that came of this parting was the background radiation that can now be found routinely in every nook and cranny of the universe. For mathematicians in pursuit of the cosmic formula this is fascinating, to ordinary mortals it says absolutely nothing at all. Even if we could lift the veil that shrouds the first 500,000 years of creation, there would be nothing for us to see but ghostly sphere formulae and monstrous numbers. Man, however, does not live by Big-Bang formulae alone. His wants are quite simple, quite naïve: he wants to know who set off this Big Bang. With nothing but numbers to keep him company, he feels alone in the universe.

The German dramatist Bertolt Brecht (1898-1956) in his *Galileo Galilei* tells us that mankind used to laugh with delight at the emptiness of the heavens. Since then we have found that the universe is teeming with numbers. Numbers are nothing to laugh about. It is just that every now and then they drive us mad.

Bohr

Numbers and Matter

Few if any scientific theories have produced such spectacular and long-lived consequences as quantum theory, which lets us achieve Faust's desire to know "whatever holds / the world together in its inmost folds". Not a single experiment has been undertaken so far that would justify questioning the validity of this fundamental theory. Physicists of the stature of Albert Einstein and Erwin Schrödinger have refused to concede that quantum theory deserves the status of an independent theory complete in itself,

127 Niels Bohr.

128 Quantum theory is the basis of modern chemistry.

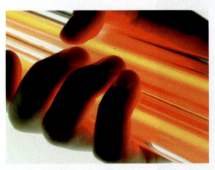

129 Quantum theory is the foundation of modern electronics.

130 Quantum theory describes the genesis of light.

preferring to think of it as a useful bundle of tentative hypotheses in need of further clarification. However, all experience to date seems to point the other way, showing that Einstein's and Schrödinger's skepticism was unjustified.

Quantum theory states the reasons for the existence of the different chemical elements and allows us to understand the structures underlying chemical reactions. It forms the basis of the whole of modern chemistry.

Quantum theory explains the way in which matter behaves if it is exposed to electric or magnetic fields. At the same time, it enables us to interpret the processes responsible for the formation of such fields. Modern electronics is inconceivable without quantum theory.

Quantum theory unravels the mystery of how light is generated—both the *warm* light of the thermal radiation emitted by white-hot bodies such as the sun or the glowing wire filament of a light bulb and the *cold* light of the laser beam. The latter, incidentally, is itself an invention made possible exclusively by quantum theory, which means that we can also thank quantum theory for such everyday applications as CD players.

The most intriguing mystery about quantum theory is that it is not understood in any full sense of the word even today, even though thousands of experiments and countless technological applications attest to its theoretical validity. What we do know about quantum theory is its principles and how to apply them. What we do not know is how these principles originate.

To express the situation through what is admittedly a very simplistic analogy, we could think of chess. What we do know about chess is the object of the game and how individual pieces are moved on the chessboard. What we do not know is how the permitted moves of the individual pieces originated. Why can the queen move any number of spaces horizontally, vertically or diagonally? Why is the king's capacity similar, but restricted to only one space in any one move? Why is the knight entitled to jump over other pieces of either color? In the case of chess, we have some sort of answer; we can cite a cleverly devised convention that is, in the final analysis, arbitrary in nature. This, of course, is the point where chess and quantum theory part company. Quantum theory has not been conjured up out of the blue by some inventive physicist; it just happens to be the only possible explanation for the way in which matter behaves.

The question is, how did the inventive physicist who first grasped the principles of quantum theory succeed in discovering a theory of matter without being at all sure what it was he had discovered?

The answer to this question focuses on one of the most outstanding scientists of the twentieth century: Dane Niels Bohr (1885–1962), who has every right to be mentioned in the same breath as Newton and Einstein. It is to his intuitive, almost somnambulist skill of teasing out creation's secrets that science owes the slow and painful birth of quantum theory. However, to appreciate how remarkable a genie he coaxed from its lamp, we must first give some thought to the dead end that was threatening physics before Bohr came along.

At the time of Niels Bohr's revolutionary innovation, physics was dominated by the belief that it was possible to understand the world through models, a belief enhanced by the clockwork metaphors used to describe how the world functions. An understanding of the world through models did not just encompass the apparently simple mechanics of pressure and impact; it also seemed to be reinforced by the kinetic theory of gases. Gases were thought to consist of minute molecules smaller than anything visible either to the unaided eye or under the microscope; these molecules were believed to be moving freely through space, rather like tennis balls. The temperature of any gas was held to be in direct correlation to the mean velocity of these molecules. It followed that the hotter the gas, the greater not only the

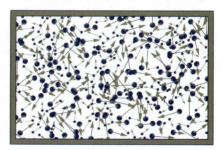

131 Schematic model of a gas in a container.

132 Fully functional toy steam engine for children dating from the mid-nineteenth century.

133 Robert Boyle.

average velocity of the molecules but also the momentum at which the molecules collide with the walls of any container that surrounds them on all sides. This, in turn, increased the pressure that the heated gas exerted on the walls of such a container—a phenomenon that provides the basis for the functioning of a steam engine.

The hypothesis that matter—not just gaseous matter, but liquid and solid matter as well—consists of molecules led to phenomenal successes in nineteenth-century chemistry. In 1808, the English physicist and chemist John Dalton (1766–1844) studied the hypothesis of two ancient philosophers, Leucippus (fl. fifth century BC) and his pupil Democritus (460–370 BC), by which all matter consists of minuscule particles—*átomoi*—which were, in the literal meaning of the Greek, physically indivisible. Dalton hypothesized that matter consisted of *molecules*, which were congregations of such *atoms*, and that these *atoms* constituted the smallest order of particles and could not be further divided by chemical (or any other) means.

If the molecules of a substance consisted exclusively of atoms of the same type, they would constitute an *element*, a term coined in 1661 by the Anglo-Irish natural philosopher Sir Robert Boyle (1627–1691). Elements are the chemical counterparts of prime numbers. Just as prime numbers are not the products of other numbers, chemical elements do not have any admixture of other substances. They are, in Boyle's words, "certain, primitive and simple, or perfectly unmingled bodies". Twelve such elements were known to him: carbon, sulphur, iron, copper, arsenic, silver, tin, antimony, gold, mercury, lead and bismuth. Very soon, the gases hydrogen, nitrogen, oxygen and chlorine; the liquid bromine; and the solid substances sodium, calcium and chrome were added to the list. The nature of many of these elements had been studied over centuries by alchemists. One could argue that the fact that none of their experiments, no matter how ingeniously devised, had succeeded in turning any of the baser metals into gold was itself proof that gold had to be an element.

Just as most numbers are not prime numbers on the model of 5, 17 or 1,601 but are composites such as $6 = 2 \cdot 3$, $18 = 2 \cdot 3 \cdot 3$ or $1602 = 2 \cdot 3 \cdot 3 \cdot 89$, most substances are not chemical elements but chemical compounds. Cinnabar (red mercuric sulphide) separates out when heated to give mercury and sulphur.[99] Water was shown to separate into two gases, hydrogen and oxygen, in an elegant experiment performed by the French scientist Antoine-Laurent Lavoisier (1743–1794), who is generally regarded as the father of modern chemistry. As a volume of hydrogen exactly twice as large as a given volume of oxygen is required to convert the latter back to water, it was safe to assume that a water molecule consists of two atoms of hydrogen and one atom of oxygen.

Although molecules could not be seen—the same obviously applies all the more to atoms—analytical chemists soon became used to the idea of these minuscule particles being the tiny building blocks of the entire material world.

Physics contributed its share to the elaboration of Dalton's atomic hypothesis. Experiments involving electricity conducted in the last quarter of the nineteenth century extended it to electrical charges. The Irish scientist George Johnstone Stoney (1826–1911) discovered in 1874, as he split water with the help of electricity into hydrogen and oxygen, that electricity seemed to have its own atomic structure—the amount of electrical charge needed to release a gram of hydrogen from water appeared to be consistent. However, it was not until 1891 that Stoney presented his thoughts to the public and proposed the term *electron* for the smallest unit of electrical charge.

At the same time, Sir Joseph John Thomson (1856–1940), the director of the Cavendish Laboratory at Cambridge, set off in a completely different direction. His research led him to prove successfully the existence of electrons. He fused *electrodes*, metal prongs, to a 20-inch-long glass tube and applied a voltage of 10,000 volts, initially with no noticeable result. When Thomson extracted the gas from inside the tube, flashes of light began to show in the developing vacuum until, when the gas was almost completely exhausted, an even flow of pale, green-tinged light appeared. The light rays that cause this fluorescence in the glass tube are emitted by the negatively charged electrode, known as the *cathode*, and are called *cathode rays*. The fact that a magnetic field applied to the glass tube makes them change direction proves them to be streams of electrically charged particles or, in other words, *electrons*. Highly sophisticated experimental designs enabled Thomson and

134 Sir Joseph John Thomson.

135 Electromagnetic fields deflect cathode rays: they consist of a stream of electrons.

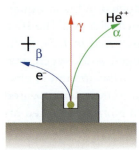

136 Radioactive substances emit radiation that an electromagnetic field will split up into alpha, beta and gamma rays.

137 Sir Ernest Rutherford.

other physicists doing research in the same area to calculate not only the electrical charge of a single electron but also its mass. They were also able to determine that electrons in cathode rays travel at one-fifth the speed of light. Finally, Thomson proved that hydrogen atoms, although they remained indivisible for the analytical chemist, could be seen from the physicist's point of view to consist of two parts: the electron and an equally large positive charge, almost 2,000 times heavier than an electron.

Further proof for the atomic structure of matter came with the discovery of radioactivity by French physicist Henri Becquerel (1852–1908) and his compatriots Pierre Curie (1859–1906) and his Polish-born wife Marie (1867–1934). Radioactive elements such as uranium, polonium and radium—these last two elements were actually discovered by Marie Curie—show three distinct emissions:

Streams of light, known as *gamma rays*.

Streams of negatively charged particles, known as *beta rays* or *beta particles*. Beta rays proved to consist of those electrons whose existence Stoney had inferred from chemical analyses and Thomson from experiments involving electricity.

Streams of positively charged particles, know as *alpha particles*. Alpha particles turned out to be four times as heavy and twice as electrically charged as the positive elementary charges that form part of the hydrogen atom. Later, it was discovered that a stream of alpha particles consists of the positively charged parts of helium atoms, the second-lightest atoms after hydrogen. Alpha and beta particles are subject to spatial deflection and segregation through magnets.

In 1911, the versatile New Zealand-born scientist Ernest Rutherford (1871–1937) fired alpha

138 Uranium atom according to Rutherford: electrons circle around the positively charged nucleus the way that the planets revolve around the sun.

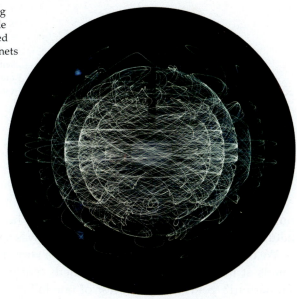

particles at a gold foil. He found that most of the alpha particles passed straight through this foil, in spite of the fact that gold foil is thousands of layers of atoms thick. Only a very small number of alpha particles—something like one in every 100,000—changed course slightly, but a tiny number remained that were violently deflected or even bounced back as if they had hit something solid.

From these results, Rutherford proposed that the atom, which has a diameter of barely a hundred millionth of a centimeter, consisted largely of empty space. A tiny part of it, with a diameter approximately one tenthousandth of the atom, is occupied by a solid center, which Rutherford called its *nucleus*. This nucleus accounts for practically the whole mass of the atom and its entire positive charge. Electrons, which revolve through the empty spaces around the nucleus, provide the compensatory negative charge. The atom, according to Rutherford's experimental results, was like a tiny solar system with the nucleus as the sun, circled by electrons in their role of planets. For Rutherford, the differences among chemical elements were accounted for simply by the number of electrons that circle their respective nuclei in each element's "solar system". This number was identical to what we call the *atomic number* in the periodic table of chemical elements. A hydrogen atom had one electron revolving around its nucleus; helium, two, lithium, three; beryllium, four; boron, five; carbon, six; nitrogen, seven; oxygen, eight; fluorine, nine; and so on up to uranium (atomic

139 Schematic depiction of Rutherford's experiment: the vast majority of alpha particles penetrate the gold foil without deflection; only a few are deflected, some of them violently.

number 92), whose 92 electrons made it the heaviest element found in nature (this last part of Rutherford's theory was soon revised).

We can see that the atom had assumed a very strange shape in scientific minds at the beginning of the twentieth century. If we were to enlarge Rutherford's gold foil by a factor of one billion (1,000,000,000), it would become a crust of gold thick enough to cover the Eiffel tower twice over and extensive enough to cover the whole continent of Europe. This crust would consist of gold atoms the size of balloons more than half a meter in diameter. The entire mass of each atom would be contained in its nucleus—a nucleus one-twentieth of a millimeter in diameter, tinier than a grain of sand. The 79 high-speed electrons (79 being the atomic number of gold) would orbit this nucleus in paths taking them all the way out to the skin of the balloon.

Persuasive though many details of this picture of the atom as a miniature solar system may be, the insurmountable difficulties and contradictions of such a model were soon revealed.

The easiest to negate was the claim that Dalton's chemically indivisible atoms are yet divisible physically; after all, each atom consists of negatively charged electrons and the positively charged nucleus.[100] It is possible that the term *atom* with its implication of indivisibility was temporarily applied to the wrong object, but the misnomer could easily be rectified by saying that it had been elementary particles such as electrons[101] which Leucippus and Democritus had anticipated with their *átomoi*.

A much more disquieting aspect was indicated because the atomic hypothesis proposed by Leucippus and Democritus had to include light. Light also consists of particles, particles that we now call *photons*. It was none other than Albert Einstein (1879–1955) who coined the term *photon* to account for the following experimental finding: shining high frequency ultraviolet light on to a zinc plate dislodges electrons from the plate;

140 The principle of the photo-electronic effect: even weak ultraviolet light dislocates electrons from an irradiated zinc slab; red light, no matter how intense it is, cannot dislocate even a single electron.

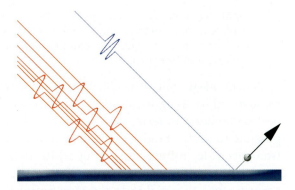

carrying out a similar experiment using visible light produces no such result.[102]

Einstein demonstrated convincingly that the dislodging of the zinc electrons can only be explained by making use of the idea of photons. A single photon will contain more energy the higher the frequency of its (parent) light. In simple language, this means that low-frequency red photons will have less energy than high-frequency blue ones. If we multiply the frequency of any given light by a constant (known as *Planck's constant* after German physicist Max Planck (1858–1947), who discovered it in 1900), the result is the energy unit represented by each photon of that light. Furthermore, it takes only one photon sufficiently rich in energy to dislodge one electron from the zinc plate. For this reason, even the weakest ultraviolet light is powerful enough to trigger a photoelectric effect, whereas visible light, be it ever so intense, will not dislodge a single electron from its zinc bed. This is because not one of the countless low-frequency photons in even an intense stream of visible light is sufficiently rich in energy to dislodge even a single electron.

Einstein's photon hypothesis caused such a stir because the idea that light consisted of particles was a kind of *revenant* in the world of science. Formulated in the seventeenth century by the great Newton himself, it was commonly held to have been refuted conclusively by a great number of nineteenth-century experiments. Received wisdom could be summarized as follows:

> Light exists as a series of single waves, not as a shower of particles.
>
> These waves were brought about by energy in the form of electrical and magnetic fields that vibrated at right angles to the direction of the wave and at right angles to each other.
>
> This was true for all electromagnetic waves ranging from low-frequency radio waves through infrared thermal radiation and visible light to ultraviolet radiation and, finally, to X-rays and gamma rays.

Light was seen as similar in form to sound. There are no sound particles, each contributing its mite of noise—sound is created by the transmission through a medium of a pressure wave triggered by vibrations at its source.

We can easily empathize with this *received wisdom* today. If light is not a phenomenon best understood in terms of a wave, how is it possible to understand why a CD, when seen from the side, refracts incoming white light into all the colors of the rainbow? It makes perfect sense to see this as the superimposition of light waves one upon another as they are reflected from neighboring grooves, which results in certain colors (light at different wavelengths) reinforcing one another, whereas others cancel each other out. However, how can photons be made to fit in with the wave theory of light, which has such weight of evidence to support it?

141 The rainbow effect on a CD is easy to understand.

An even greater embarrassment was the realization that the negative charge of electrons oscillates as they circle around the nucleus so that they function as transmitters of electromagnetic waves. In other words, each electron transmits to the outside world light whose frequency corresponds to its revolutions around the nucleus. Emitting these waves entails for each electron a continuous drag on the energy it requires to maintain its circular motion. This

142 Neighboring grooves reflect white light (which is a mix of all colors) in such a way that some colors (wavelengths) are enhanced and become visible, whereas others cancel each other out.

drag should eventually result in the electron crashing into the nucleus, which in turn would spell the end for its atom. Yet the universe does not implode within milliseconds as this scenario would demand; on the contrary, it has been going from strength to strength for billions of years. How can we explain this if our model of the atom as a miniature solar system is correct?

Rutherford's model also failed to address another mystery: why the chemical properties of elements should be so radically dependent on their atomic numbers, given that these were no more than the number of electrons circling their nuclei. Why do chlorine atoms, whose atomic number is 17, leave their signature in the shape of a poisonous, greenish-yellowish gas, whereas argon atoms, only one step further on at atomic number

18, express themselves in a harmless "noble" gas, which treats practically all possibilities of chemical reaction with disdain? Their next-door neighbor, potassium (at number 19), is a metallic solid, which causes powerful explosions when even small amounts of it are brought into contact with water. Our solar system would not be altered significantly if the sun were to lose a dwarf planet such as distant Eris—yet for atoms, a difference of one electron either way means an entirely different set of properties. What could be the underlying reasons for this?

Finally, the very notion of *átomoi*, of indivisible particles, was philosophically suspect. Plato had already pointed out the contradictory character of Democritus' ideas. If atoms are to have different sizes and shapes or, rather, if they are to have sizes and shapes at all, why should they then be indivisible? The volume represented by a Democritean atom is surely amenable to further downscaling; for instance, we can at least think of it as having upper and lower halves. What guarantee is there that the two will always stay together? Equally, consider a version of atomic theory that takes a slightly different tack from Democritus' original: if atoms are mere points and therefore without extension, how do they manage to fill space?

Bohr realized that the solution for all these problems would have to take the form of a radical break from the alluring models that physicists had so lovingly constructed. Neither promising analogies, like the one between the structure of the atom and the solar system, nor vague philosophical hypotheses about the indivisibility of elementary particles, such as the Democritean atomic theory, were at all likely to prove helpful in the formulation of a comprehensive theory of matter. To put it more succinctly, it is counterproductive to think of an electron as a billiard ball, only infinitely smaller, if what we want to know is actually what electrons contribute to the structure of matter in all its various forms including, and in a prominent place, billiard balls. In the end, over-explicit models merely take us around in circles.

It may be easy to discard over-explicit models; however, it is not so easy to find a replacement for them. How can anyone even start looking for a replacement without some sort of template to follow? How is it possible to initiate a theory of matter and still observe the vital commandment that forbids creating a graven image of it?

Bohr's ingenious discovery drew on an imageless theory of matter, which dated back to 1885 and to the extraordinary efforts of a Swiss elementary-school master, Johann Jakob Balmer (1825–1898). Again, we need to glance at the history leading up to this decisive moment to gain a proper understanding of the achievements involved.

| 2,518,130 | 2,437,290 | 2,303,240 | 2,056,410 | 1,523,310 |

143 Schematic depiction of the hydrogen spectrum with exactly measured wave numbers (per meter).

In 1859, the German chemist Robert Bunsen (1811–1899) and his compatriot, the physicist Gustav Kirchhoff (1824–1887), carried out a series of highly sophisticated spectroscopic experiments and made the following discoveries:

> Chemical elements, when heated to incandescence with a Bunsen burner, give off highly characteristic colors.

> These colors are unique to each element.

> Refining the experiments using electricity as a replacement for thermal energy and at the same time making use of a gas-discharge tube gives better results, demonstrating that the colors show up more distinctly the higher the temperature and the more rarefied the gaseous element.

Building on these discoveries, Bunsen and Kirchhoff took their research further:

> They produced lists with exact descriptions of where on the spectrum each element left its characteristic signature.

> They even succeeded in discovering new elements through spectroscopy such as cesium and rubidium, both of which have a characteristic violet signature.

> They devoted extra attention to the light emitted by hydrogen, the lightest element, with the atomic number one.

Their optical measurements showed that incandescent hydrogen emits light with the wave numbers 1,523,310; 2,056,410; 2,303,240; 2,437,290; and 2,518,130. These numbers state how often light of that specific color oscillates back and forth within a distance of one meter. The shorter the wavelength of light, the farther it is shifted toward the blue end of the spectrum and the higher its frequency. The wave number is proportional to frequency.

Bunsen and Kirchhoff measured the characteristic lines of hydrogen as follows:

144 Robert Bunsen and Gustav Kirchhoff.

long-wave red with a wave number of 1,523,310,
turquoise with a wave number 2,056,410,
blue with a wave number 2,303,240,
indigo with a wave number 2,437,290 and
short-wave violet with a wave number of 2,518,130.

These are the results of remarkably precise optical measurements, although the width of the spectral lines is still enough to introduce a certain margin of error. We should take into account that the tens digits of the numbers above—1, 1, 4, 9, 3—cannot be relied on absolutely and that the units digits, shown above as zeroes, are possibly below the measurement threshold. What remains is a set of values measured to five significant figures—still a formidable degree of accuracy. In spite of the existence of such reliable data, however, it was still a long time before anyone discovered a mathematical law governing the wave numbers of the spectral lines of hydrogen in their distribution from red to violet.

This is where our hero, Johann Jakob Balmer, the unassuming, God-fearing Swiss elementary-school teacher, entered the stage in 1885. What is remarkable about him is not just that he succeeded in correlating the wave numbers where others had failed; more important for the future of science was the method that he used. Whereas his predecessors had worked with model-based methods, Balmer based his inquiry solely and squarely on the five data available to him. These and his unshakable conviction that he was going to solve the problem were sufficient for him. Here is what he did.

In a first step, he calculated the ratio between the second smallest number in the series and the smallest, turquoise to red:

$$2,056,410:1,523,310 = 1.349962.$$

He then rounded 1.349962 to 1.35, which is perfectly legitimate given the margin of error in the measurements. Rounding 1.349962 to five significant figures yields 1.3500. Since 1.35 = (135/100) = (27/20), Balmer assumed that the exact ratio between the wave numbers of turquoise and red was 27:20.

His assumption that he was justified in replacing the complex fraction[103] 2,056,410/1,523,310 with the simple 27/20 reflects his underlying belief that nature's laws could be relied on to be simple.

Next, he took the ratio of the third smallest to the smallest, blue to red:

$$2,303,240:1,523,310 = 1.511997.$$

He rounded this 1.511997 to 1.512, which is again perfectly legitimate. Rounding 1.511997 to five significant figures yields 1.5120.

Since 1.512 = 1,512/1,000 = 189/125, Balmer assumed the exact ratio between the wave numbers of blue and red to be 189:125.

The result obtained in his third foray: fourth smallest to smallest, indigo to red,

$$2,437,290:1,523,310 = 1.59996,$$

Balmer truncated to a straightforward 1.6.

Since 1.6 = 16/10 = 8/5, he assumed the exact ratio between the wave numbers of indigo and red to be 8:5.

In his fourth and last foray, largest to smallest, violet to red, he obtained this result:

$$2,518,130:1,523,310 = 1.653064.$$

This was a tricky one, because it was not immediately obvious that this complex ratio could be simplified by sufficiently accurate rounding. However, Balmer persisted in his belief that a simple, and therefore exact, ratio was hidden somewhere in this numerical haystack. After all, that belief had already been justified by his earlier results. He began to transform the fraction in a series of steps. He applied a method that was familiar to every elementary-school teacher in his day. It was, moreover, a time-hallowed one, which had already been applied with consummate skill by the Pythagoreans.[104]

The first step was to remove the units digit:

$$2,518,130:1,523,310 = 251,813:152,331.$$

Next, he calculated the nearest whole-number value of the ratio: $251,813:152,331 \approx 2$. Then he calculated the difference between his original ratio and this whole number:

$$2 - (251,813:152,331) = 52,849:152,331.$$

He formed the reciprocal of the ratio of this difference: 52,849:152,331. He did this because the reciprocal would enable him again to calculate the nearest whole-number value—obviously 3—and to find the difference as before:

$$3 - (152,331:52,849) = 6,216:52,849.$$

The next step again consists in forming the reciprocal of the difference, which gave him the calculation:

$$9 - (52,849:6,216) = 3,095:6,216.$$

The reciprocal of this last ratio, 6,216:3,095 = 2.008400, corresponds almost exactly to 2. It follows that 3,095:6,216 is almost exactly 1:2.

Balmer therefore assumed that the exact ratio between the wave numbers of violet to red light would be obtained from the *cascade* of

$$9 - (1:2) = 17:2$$
$$3 - (2:17) = 49:17$$
$$2 - (17:49) = 81:49.$$

The two values

$$2,518,130:1,523,310 = 1.653064\ldots \text{ and } 81:49 = 1.653061\ldots$$

are identical to six significant figures. The interim result to which Balmer had been guided by his conviction that nature's laws were nothing if not simple consisted in his calculation of the four ratios—which he took to be exact—

turquoise to red:	27:20
blue to red:	189:125
indigo to red:	8:5
violet to red:	81:49

These ratios state by how much the frequencies of the turquoise, blue, indigo and violet lines of incandescent hydrogen are higher than its red component.

But what *law* follows from these numbers?

Contemplating the denominators 20, 125, 5 and 49, Balmer found it remarkable that the first three are divisible by 5 and that, after carrying out the division by 5, he was left in each case with a square number: 4, 25 and 1. The fourth number, 49, was itself a square number. He thought it unlikely that chance accounted for the fact that 25 was the fifth square and 49 the seventh. That the sixth square, 36, did not appear where he would have expected it—namely, in the frequency ratio of indigo and red—Balmer attributed to its having disappeared in the process of *canceling*. The same reason may account for the second square, 4, featuring in the first ratio instead of the expected fourth square, 16.

Balmer conjectured quite rightly that these irregularities could be corrected if he reconfigured the red-light wave number of 1,523,310, to which he had related all other wave numbers. This he did by introducing the multiplier 5/36 and so arriving at an "absolute" wave number. The significance of 5/36 will be readily understood once we realize that, by virtue of it, the factor 5 is removed from the preceding denominators while, at the same time, the missing square, 36, is introduced. This "absolute" wave number, of course, is easily calculated as

$$1,523,310 \cdot 36/5 = 10,967,832.$$

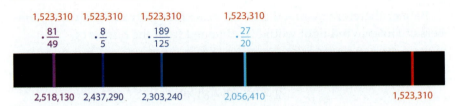

1,523,310	1,523,310	1,523,310		1,523,310
$\cdot \dfrac{81}{49}$	$\cdot \dfrac{8}{5}$	$\cdot \dfrac{189}{125}$		$\cdot \dfrac{27}{20}$

| 2,518,130 | 2,437,290 | 2,303,240 | 2,056,410 | 1,523,310 |

145 Balmer obtained the wave numbers listed below the hydrogen spectrum by multiplying the wave number of red light by the pertinent fractions.

The value of this absolute number is so complex because all numbers connected to waves are related to the *meter* as the unit of length measurement. They state how many wave trains of the light frequency under discussion can be accommodated within one meter. When, in 1790, the French *Assemblée Nationale* created a committee to formulate the definition for the new length measurement, the globe was chosen as the frame of reference for what we might call the *ur-meter*. By this method, one meter was to be the ten-millionth part of the quarter circle extending from the equator to the North Pole via Paris. However, from Balmer's point of view, it would have made much more sense to make the unit of length measurement hinge on the hydrogen atom. This would have given us a *meter* that would have altered Bunsen and Kirchhoff's original measurements and Balmer's subsequent calculations so that the final result — 10,967,832 — would have been exactly 10 million. Sadly, this fact was unknown to the *Assemblée Nationale* back in 1790. Be that as it may, the "absolute" wave number 10,967,832 has since been called the *Rydberg constant* after the Swedish physicist Janne Rydberg (1854–1919).

Balmer went on to convert all five frequencies of the hydrogen lines into multiples of this *Rydberg constant*, and we will study these multiples next.

To define the wave number of red light, Balmer fixed the ratio

$$5{:}36.$$

To define the wave number of turquoise light, higher than that of red light by the ratio of 27:20, Balmer calculated

$$(27{:}20) \cdot (5{:}36) = 3{:}16.$$

To define the wave number of blue light, higher than that of red light by the ratio of 189:125, Balmer calculated

$$(189{:}125) \cdot (5{:}36) = 21{:}100.$$

To define the wave number of indigo light, higher than that of red light by the ratio of 8:5, Balmer calculated

$$(8:5) \cdot (5:36) = 2:9.$$

To define the wave number of violet light, higher than that of red light by the ratio of 81:49, Balmer calculated

$$(81:49) \cdot (5:36) = 45:196.$$

With these results,

$$5:36, 3:16, 21:100, 2:9 \text{ and } 45:196,$$

Balmer was nearly there. The denominators he had obtained were all square numbers; 16 and 9 were the only ones not to fit into a regular ascending sequence.

His solution was to replace 16 with 64 =16 · 4 and 9 with 144 = 9 · 16. This left him with

$$36, 64, 100, 144 \text{ and } 196,$$

the sixth, eighth, tenth, twelfth and fourteenth squares in regular ascent. All that was needed now was to expand the second ratio—3:16—by 4 and the fourth—2:9—by 16. The result now contained only factors whose multiplication by the Rydberg constant produced the exact wave numbers of incandescent hydrogen:

$$5:36, 12:64, 21:100, 32:144 \text{ and } 45:196.$$

Balmer made sure, moreover, that all denominators divided by 4 continue to remain squares—9, 16, 25, 36 and 49:

$$36 = 9 \cdot 4, 64 = 16 \cdot 4, 100 = 25 \cdot 4, 144 = 36 \cdot 4 \text{ and } 196 = 49 \cdot 4$$

At the same time, he had demystified the numerators. He simply swapped the multiplication sign for the minus sign to obtain the following results:

$$5 = 9 - 4, 12 = 16 - 4, 21 = 25 - 4, 32 = 36 - 4 \text{ and } 45 = 49 - 4.$$

With this brilliant method (which calls to mind the discussion earlier in the book about number magic), Balmer had discovered the formula for the wave numbers of hydrogen:

Let the Rydberg constant 10,967,832 be multiplied by the following factors:

$$\frac{9-4}{9 \cdot 4}, \quad \frac{16-4}{16 \cdot 4}, \quad \frac{25-4}{25 \cdot 4}, \quad \frac{36-4}{36 \cdot 4}, \quad \frac{49-4}{49 \cdot 4}.$$

Bohr realized that Balmer's formula gave physics the chance to discover the *true* theory of matter or, at the very least, of hydrogen. The point was that Balmer had arrived at his law by a straightforward conversion of data into *numbers*, which he conjectured to be accurate. He had used neither a theory that might require adjustment nor a model drawn from nature.

In other words, by grounding his investigation solely on his own highly sophisticated technique and on his unshakable belief that the laws of nature were characterized by simplicity, Balmer had managed, in the phrase later coined by Einstein, to "sneak a look at God's playing cards".

What remained to be done was to put into words what Balmer had seen in "God's playing cards" and, in doing so, to resist the temptation to read more into them than was actually there. In 1913, Bohr published an epoch-making article, in which he demonstrated how the hydrogen atom could be calculated on the basis of Balmer's formula. Bohr's starting point was the realization that the fractions of the formula can be converted into differences:

$$\frac{9-4}{9\cdot4} = \frac{1}{4} - \frac{1}{9},$$

$$\frac{16-4}{16\cdot4} = \frac{1}{4} - \frac{1}{16},$$

$$\frac{25-4}{25\cdot4} = \frac{1}{4} - \frac{1}{25},$$

$$\frac{36-4}{36\cdot4} = \frac{1}{4} - \frac{1}{36},$$

$$\frac{49-4}{49\cdot4} = \frac{1}{4} - \frac{1}{49}.$$

What is the significance of these individual fractions, all of them reciprocals of squares, in terms of physics?

Bohr offers the following explanation: the hydrogen atom is characterized by a number of stable states, which he called *quantum states*. Each state has its corresponding number: 1, 2, 3, 4, 5, 6, 7, ..., its *quantum number*. In the state characterized by the quantum number 1, the electron is bonded to the atom with a precisely defined amount of energy. This amount of energy is precisely that of one photon of light whose wave number is identical with the Rydberg constant of 10,967,832. As quantum numbers become larger, the energy by which the electron is bonded to the atom decreases in keeping with the preceding fractions. Therefore, the bonding

146 Balmer obtained the wave numbers listed below the hydrogen spectrum by multiplying Rydberg's constant (notated in black) by the pertinent fractions. For this operation, Balmer obtained the fractions' numerators and denominators—his method is summarized below the fractions—by the application of an almost miraculous law.

energies are, in accordance with the quantum number series 2, 3, 4, 5, 6, 7, ..., lower by 1/4, 1/9, 1/16, 1/25, 1/36, 1/49, ... than are those pertaining to the state with the quantum number 1. If the hydrogen atom switches from an *excited* state with a high quantum number to a less excited state with a low quantum number, the difference between the two amounts of bonding energy is identical to the energy of the photon that the hydrogen atom emits as light. In the visible-light range, this is particularly relevant for the transition from states with the quantum numbers 3, 4, 5, 6, 7, ... to the state with the quantum number 2. The energy of the photons emitted on these occasions is therefore a multiple 1/4 – 1/9, 1/4 – 1/16, 1/4 – 1/25, ... of the bonding energy of the state with quantum number 1. This is directly in keeping with Balmer's formula.[105]

It goes without saying that other transitions from higher to lower quantum states are possible and that they actually take place. The following table lists the factors that, after multiplication by Rydberg's constant, specify all possible wave numbers of incandescent hydrogen:

$$1 - \frac{1}{2^2}$$

$$1 - \frac{1}{3^2} \qquad \frac{1}{2^2} - \frac{1}{3^2}$$

$$1 - \frac{1}{4^2} \qquad \frac{1}{2^2} - \frac{1}{4^2} \qquad \frac{1}{3^2} - \frac{1}{4^2}$$

$$1 - \frac{1}{5^2} \qquad \frac{1}{2^2} - \frac{1}{5^2} \qquad \frac{1}{3^2} - \frac{1}{5^2} \qquad \frac{1}{4^2} - \frac{1}{5^2} \quad \cdots$$

$$1 - \frac{1}{6^2} \qquad \frac{1}{2^2} - \frac{1}{6^2} \qquad \frac{1}{3^2} - \frac{1}{6^2} \qquad \frac{1}{4^2} - \frac{1}{6^2} \quad \cdots$$

$$1 - \frac{1}{7^2} \qquad \frac{1}{2^2} - \frac{1}{7^2} \qquad \frac{1}{3^2} - \frac{1}{7^2} \qquad \frac{1}{4^2} - \frac{1}{7^2}$$

$$\vdots \qquad\qquad \vdots \qquad\qquad \vdots \qquad\qquad \vdots \qquad\qquad \ddots$$

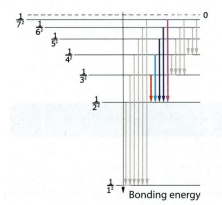

$\frac{1}{7^2}$
$\frac{1}{6^2}$
$\frac{1}{5^2}$
$\frac{1}{4^2}$
$\frac{1}{3^2}$
$\frac{1}{2^2}$
$\frac{1}{1^2}$

0

Bonding energy

147 The light of the hydrogen atom is generated by the electron "jumping" from a higher quantum state to a lower one—an explicit visualization of this *quantum leap* is neither necessary nor meaningful.

The first column of the table on the previous page, the so-called Lyman series, contains the wave numbers of highly energized ultraviolet light, which was discovered in 1906 by American physicist Theodore Lyman (1874–1954). Here the electron drops back to the state with the quantum number 1. Columns 3 and 4 contain the wave numbers of low-energy infrared light, which were registered shortly afterward by Friedrich Paschen (1865–1947) of Germany and Frederick Sumner Brackett (1896–1972) of the United States. In them, the electron drops back to states with the quantum numbers 3 and 4. The second column is the Balmer series of incandescent hydrogen in visible light, which was our starting point.

What is striking is the terminology that Bohr is careful not to use—the electron is not called a particle, in the sense of a microscopically small planet completing its orbits around the nucleus. Bohr is already aware that no purpose is served by referring to the location or the velocity of the electron as part of its atom. The far too descriptive picture of the electron as a "tiny little ball" is misleading and is an impediment to our understanding of the atom. The state of an electron within an atom is defined by its quantum *number*.[106]

What Bohr succeeded in doing for the hydrogen atom was expanded by his students and collaborators, such as Werner Heisenberg, Austrian physicist Wolfgang Pauli (1900–1958), English physicist Paul Dirac (1902–1984), American chemist Linus Pauling (1901–1994), the skeptical Erwin Schrödinger and many other Nobel Prize winners, into a complex mathematical theory applicable not only to the atoms of all chemical elements but also to molecules, crystals, and ultimately to our whole material world. The complexity of our world is such that a single number is insufficient to represent a quantum state.[107] It became necessary to specify a state according to a rule-based system of quantum numbers. This is not the place to discuss the details, for which the reader is referred to textbooks on quantum physics. What matters to us is the underlying principle of this theory that numbers, alone and unsupported, are the building blocks of creation.

American physicist Victor Weisskopf (1908–2002), another of Niels Bohr's well-known students, gave an instance of this principle at work in a lecture on quantum theory:

148 Victor F. Weiskopf.

> Once the number of electrons is known, it is possible to deduce the properties of the atom from the quantum states of its electrons. We got first-hand experience of this when the element plutonium was isolated for the first time from nuclear reactions. (With the exception of a few traces in naturally irradiated uranium ore, plutonium is not found in nature, as it disintegrates after approximately 40,000 years.)
>
> The first amounts of plutonium obtained were so extremely small that no observation of its properties was feasible. Nevertheless it was clear that in a plutonium atom 94 electrons were bonded to the nucleus. This enabled us to calculate the properties of plutonium from the quantum states of those electrons:
>
> - It was sure to be a metal.
>
> - One cubic centimeter was likely to weigh about 20 grams.
>
> - It was probably silvery colored with a brown tarnish.
>
> We were also able to calculate its electrical and thermal conductivity and its elasticity.
>
> All this was done on the basis of Bohr's quantum theory and the one number 94. It was a dramatic moment when finally the first cubic millimeters of plutonium were available and all these predictions were proved true. This example documents the incomprehensible power of the quantum theory of atoms: once the quantum states of electrons in any atom have been calculated, one knows how such atoms will combine to form molecules and how these molecules will form gases or liquids or solidify into crystals.

Finally, we conclude with three remarks.

First, we should mention that the development of Bohr's ideas into quantum theory explains, among other things, why the spectral lines of hydrogen studied by Balmer have a particular width (a fact, as we pointed out previously, that also introduces a margin of error into the measurements). Balmer's confident assumption that *true* integer ratios will be simple presupposes an ideal hydrogen atom that exists independently of its environment—a precondition impossible to realize even in a gas-discharge tube with extremely rarefied contents. If there were such an atom, its spectral lines would indeed be reduced to mathematically exact lines located where Balmer had predicted them—the drawback being of course that they would be impossible to measure, because mathematically exact, *infinitely narrow* lines would be invisible in the spectrum.

Arriving at perfect numbers through imperfect means recalls an anecdote with Bohr at its center, told by Werner Heisenberg:

> After the meal [in a mountain chalet] the rota of chores resulted in Niels [Bohr] having to wash the dishes. I had to clean the stove and the others were left to chop wood and make themselves useful in other ways. It goes without saying that hygienic standards in a chalet kitchen won't bear comparison with those in town. Niels commented on the situation saying: "Dishwashing is a process not dissimilar from our language [the language of physics]: we have dirty dishwater and dirty towels, and yet we manage to end up with clean glasses and clean plates."

In the version of the German physicist and philosopher Carl Friedrich von Weizsäcker (b. 1912), who was also present, the same point is made:

> Bohr surveyed his work with pride and said: "So, you can actually make dirty glasses clean using dirty water and a dirty cloth—tell that to a philosopher and he won't believe you."

Balmer was in precisely this situation. Hydrogen atoms far removed in the gas-discharge tube from the ideal state of a single atom in an empty universe and measurements riddled with errors and inaccuracies did not prevent

149 Carl Friedrich von Weizsäcker.

him from treating the atom as if it were emitting light in *ideal* conditions.

Second, the story of the birth of quantum theory flatly contradicts traditional assumptions about the genesis of scientific theories. According to these assumptions, a dominant doctrine is challenged and *falsified* by new findings. In other words, new empirical data is found that can no longer be interpreted in terms of the dominant doctrine. The revolutionary findings are then transformed into an improved version of the doctrine, the end of whose life cycle is spelled by a new round of falsification. In the case of quantum theory, it did not happen that way. There was no *dominant doctrine* about atoms, just graphic models that had proved counterproductive and incapable of improvement. Quantum theory is not a transformation of a theory that was in need of improvement. Its principles enable us to differentiate, among such terms inherited from classical physics as energy, place and velocity, between those that can meaningfully be represented by quantum numbers in a system (such as an atom) and those that cannot. That such principles were going to be available could not in the least have been predicted on the basis of any theory predating quantum physics.

Third, an interesting parallel can be drawn between the achievements of Balmer and Bohr in their foundation of quantum theory and those of Kepler and Newton in their foundation of theoretical mechanics. In a manner analogous to Balmer's work, Kepler had distilled numerical material obtained from measuring the orbit of the planet Mars into mathematical laws: the famous laws of planetary motion, which state that the planets describe elliptical, not circular, orbits around the slightly eccentrically positioned sun. Kepler understood no more than Balmer did what the laws discovered by him ultimately signified. However, he did experience them, again in parallel with Balmer, as miraculously *harmonious*—he, too, had "sneaked a look at God's playing cards". Just as Bohr validated Balmer's formula by integrating it into quantum theory, Newton grounded Kepler's mathematical laws in the principles of mechanics. Newton, however, in antithesis to what Bohr did with quantum numbers, did not relate his theory to numbers but introduced the term *force*. In the case of the planets, it takes the form of the gravitational field of the sun. *Force* is the term that underpins his mechanical principles. To this day, we do not actually know what to make of this term. It may well be that the last task waiting to be tackled in theoretical physics is the dissolution of the term *force* into that of *number*.

Pascal

Numbers and Spirit

L'homme n'est qu'un roseau, le plus faible de la nature, mais c'est un roseau pensant.

Man is only a reed, the feeblest thing in nature; but he is a thinking reed. The universe has no need to arm itself to crush him: one breath of wind, one drop of water and he is dead. But though the universe might crush him, man would still be nobler than his destroyer, for he is aware: aware of his own death and aware of the advantage the universe has over him. The universe knows nothing of this.

150 Blaise Pascal.

> Thus all our dignity lies in thought. It is through thought we must raise ourselves, and not through space and time, which we can never fill. So let us strive to think well: this is the mainspring of morality.

The French philosopher and mathematician Blaise Pascal (1623–1662) conceived this thought in full awareness of the pitiful status of human existence, which is defenseless against a cold, soulless world and its own ultimate decay. He diagnoses both grandeur and misery in human beings: "Man knows that he is wretched: therefore he is wretched, since wretched he is; but this very knowledge means that he is also great.

"Let us strive to think well": this is Pascal's message for us. It is equally instructive to note what he does *not* enjoin us to do.

First, there is no point in wanting to *conquer the universe*—how stale this phrase still seems long after it was flogged to death by journalists at the time of the moon landings. Again and again, we learn by painful experience what an illusion it is that we have complete control over even our own bodies—let alone the universe.

Second, our constant search for diversion and amusement is no more than a futile bid to deny our wretchedness. Pascal demonstrates considerable psychological insight in describing the basic situation:

> When I have occasionally set myself to consider the different distractions of men, the pains and perils to which they expose themselves at court or in war, whence arise so many quarrels, passions, bold and often bad ventures, etc., I have discovered that all their unhappiness arises from one single fact, that they cannot stay at peace in their own room. A man who has enough to live on, if he knew how to enjoy staying at home, would not leave his home to go to sea or to besiege a town. No one would go to the expense of buying a commission in

the army, if it were not insufferable to be confined to one town; and the only reason that men seek to while away their time playing games is because they cannot happily remain at home. On further consideration I have found only one explanation for all this confusion, namely, the natural poverty of our feeble and mortal condition, which is so miserable that nothing can comfort us once we have recognized it.

An anecdote told by the American mathematician Philip J. Davis (b.1923) has much the same drift:

> There was this well-known, fabulously rich financier who had made so much money that neither he nor his family nor his heirs would ever be able to spend it all. In spite of his immense wealth he continued to be active, aggressively making more and more millions; according to him, this was "the only way to escape from reality".

152 Philip J. Davis.

Where, asks Philip Davis, may we find the seat of this *reality*?

Third, it is pointless for us to seek to make a survey of the world using only our senses. Pascal does not suggest that we can somehow purify our senses and so let ourselves experience unadulterated *reality*. In fact, our senses do not deceive us: Parmenides' suspicions were unfounded, and we can forget about those well-known optical illusion graphics that are part of the stock-in-trade of the psychology of perception. Nor is it the case that our senses impart to us a faithful representation of an actually existing state of affairs by simply dumping their catch on to our *tabula rasa,* as the empiricist school would have had us believe. In fact, our senses do not lie to us; neither do they tell us the truth.[108] They simply do not deal in that kind of communication. What our senses do is to expose our consciousness to a welter of perceptions. What our *consciousness* does with this data—this *tohu bohu*[109]—has nothing to do with our *senses*. It is *thinking* that allows us to tease meaning from sensory data, using purposeful selection and elimination to distill a whole cosmos from the swarms of data that bombard us like so many mosquitoes. The impressive case studies of the English neurologist Oliver Sacks (b. 1933)—for example, his legendary *The Man Who Mistook His Wife for a Hat*—demonstrate how narrow and treacherous the path is that leads from the mere receipt of sensory data to conscious perception.

Perception presupposes thought, and it is thought that enables us to understand that sensory data are what they are. The world we experience is the result of our thought—in Pascal's words:

> By space the universe encompasses and swallows me up like an atom;
> by thought I comprehend the world.

How does this idea deal with the age-old history of the universe? For most of it, humanity is conspicuous only by its absence. We did not see the dinosaurs; we will not see the death even of our own small sun. The short answer is that it does not; the two notions are incompatible. In Pascal's terms, there is no such thing as an event without a thinking entity being there to witness it; the idea is as abstruse as a square circle.

This is why all reports concerning the creation of the world are myths. This is true of the version of the *Elder Edda*, according to which, at the beginning,

> nowhere was there earth nor heaven above / But a grinning gap and grass nowhere. To the north and to the south extended the icy and the fiery worlds of Niflheim and Muspelheim. The heat emanating from Muspelheim made the ice of Niflheim melt, and out of the drops arose the giant Ymir....

It is also true of the scenarios devised by modern cosmologists:

> One hundred thousandth of a second after the Big Bang electrons, positrons, neutrinos, antineutrinos and photons were all there was in an ultra-dense, 100 billion degree hot soup. Eleven hundredth of a second later it has cooled to 30 billion degrees, there are already some neutrons and about twice as many protons, one second later the universe has cooled to 10 billion degrees....

Is this how it all took place? An idle question—there were no witnesses. The value of such stories consists solely in their compatibility with today's world-view—to the extent that such a thing is possible and meaningful to

153 The *Ultra Deep Field* photographed by the Hubble telescope: the photo shows the deepest recesses of the universe that we have seen to date. At the same time, this is news that is more than ten billion years old. According to cosmologists, the galaxies in the photo developed only a short while after the Big Bang.

ascertain—and in how far the poetic potential of the tale[110] can stimulate our imagination.

On the far side of myth, creation reenacts itself every time a baby is born. It unfolds in all its complexity for every one of these babies as they set in motion the process of turning mere sensory data into sense.

In the same way, *Armageddon*, *Ragnarok* and *the cold death of the universe* are all more or less acceptable myths about the end of the world. Going again to the far side of myth, it is equally true that the end of the world is reenacted in the death of every individual; the death of each and every human being wipes out a whole universe.

"Let us strive," Pascal demands, *"to think well"*, for this, as he believes, is the basis of appropriate moral conduct. And what is it to "think well"? On what is good or, as we might call it, "right" thinking based? It cannot be based on congruence with sensory perception because perception, as we have seen, is subordinated to thinking. However, there is such a phenomenon as misdirected thinking, thinking running amok. This is radically different from plainly mistaken reasoning, which is susceptible to correction. Individual wrongdoing and the colossal catastrophes of world history provide evidence for the effects of fundamentally flawed thinking as well as showing how prone we are to allow ourselves to be led into labyrinths and to get lost there. One sure way for thinking to go astray is for it to become reckless.

Albert Camus showed great artistry in making reckless thinking the subject of his drama *Caligula*. The death of his beloved sister makes the young Roman emperor, Gaius Caligula, realize that people die and they are not happy: a terse insight and yet reminiscent of what Pascal says about human misery and human grandeur. In his titan-like revolt against the stupidity of the gods and against a dispensation of human affairs that he finds unbearable, Caligula decides, in his own words, to live "by

154 Albert Camus.

the light of truth", which for him amounts to "casting off all moderation". He explores the limits of his imagination with novel ways of humiliating others, with unheard-of cruelties and murders most foul;[111] he shows his contempt for those who proffer abject adulation to him by setting himself up as the goddess Venus, demanding ritual adoration; and he literally clamors for the blue color of the sky to be handed to him right up to his disconsolate end, which, as Camus writes, combines the "story of a superior suicide and the story of the most human and most tragic of errors".

To return to what sets the standard for "right" thinking, we must confess that this is too complex a question to give a comprehensive answer here. We must content ourselves with a few notes.

Some signposts that are supposed to direct thinking toward a proper course appear convincing at first sight; on closer inspection, they turn out to have arisen from how their society was thinking at the time. An example of this can be found in the philosophy of the Milanese economist and criminologist Cesare Beccaria (1738–1794). Beccaria felt that the life of any human being was worth protecting and so opposed the death penalty on principle.[112] This idea, which set the key for Beccaria's entire thinking, is broadly accepted in Europe; it is largely rejected in the United States and in a great many other countries. History has also shown that it is possible to be a fervent supporter of Beccaria's humanism without rejecting the death penalty for crimes against humanity.[113] Is it not arguable that there was every justification for the trial of Nazi genocide Adolf Eichmann (1906–1962) to end with the verdict that he had forfeited his life and deserved to be hanged?[114] What happens in this case to the maxim of the unconditional protection of human life? What happens to the maxims of pacifists when they are confronted with a war of aggression initiated by a criminal regime? Sobering reflections like these make us doubt the validity of generalized maxims of moral judgment. Are there not exceptions to all such maxims? What of Guantánamo? What about the "waterboarding" of terror suspects as part of their interrogation? Can it be argued that these are such exceptions? Is it in fact true that the cynical motto "anything goes" advanced by U.S. philosopher of science Paul Feyerabend (1924–

155 The first few bars of the second movement of Schubert's string quintet in C major D956.

1994) is the last post for critical thought? Has Pascal set us a task that is quite simply impossible? Was he wrong to discern thought as the essence of human grandeur?

Does the near-religious experience offered by the greatest art perhaps provide a way out of the quagmire of relativism? To cite a few examples completely at random—readers are invited to substitute their own tried-and-proven vehicles—does not Schubert's C major string quintet carry with it a message of transcendent, of literally *absolute* beauty? Does not Emily Dickinson speak with a kind of *absolute* depth, even though—or perhaps because—some of her lines seem to defy interpretation? "The Feet, mechanical, go round— / Of Ground, or Air, or Ought— / A Wooden way / Regardless grown, / A Quartz contentment, like a stone." And what about Turner's *Rain, Steam and Speed—The Great Western Railway?* Do the colors and shapes of this painting not transmit an *absolute* power that easily bests the most authentic, the most graphic, the most eloquent *verbal* description of a technological achievement?

We have yet to show, however, how an intuitive grasp of whatever it is that is revealed to us in works of art can be translated into landmarks along the path of "right" thinking. Dissociated from their works of art, these revelations prove impossible to put into words; the work of art unfolds its meaning only in a personal dialogue with the listener, the reader or the viewer and is beyond the reach of objective analysis.[115]

For numbers to be understood, on the other hand, neither leisure nor sensitivity is required; they do not involve us in individual dialogues but simply exist as objective data. However, like great works of art, numbers can also be experienced against an absolute background, if they are not

156 Emily Dickinson.

157 William Turner: *Rain, Steam and Speed*, painted before 1844.

158 A detail of Roman Opalka's attempt to appropriate as many numbers as possible.

regarded solely as a means for counting but also investigated with a notion to understanding their essence. They can be understood as the products of a process that includes the possibility of infinitely many numbers. It is the *infinite* that points the way for our thinking about numbers. Not surprisingly, it was Pascal, the philosopher and mathematician, who drew attention to this.

When our thinking about numbers is "right" in Pascal's sense of the word, it is not individual numbers that matter—as they would in bookkeeping and accountancy—but the *process of counting* as such. Individual numbers are fossils, an inarticulate testimony to the ongoing project[116] of counting, of beginning with one and progressing step by step from each number to the next higher one.

Counting appears deceptively simple or even banal, and the effort required to become aware of its inherent complexity is thereby so much the greater. The French artist Roman Opalka (b. 1931) contributes to this effort by interpreting continuous counting as art. With what can almost be called manic obsession, he painstakingly traces out number upon number and builds huge painted tables in finely chased script. Decades ago, he started with 1 and has been adding to it ever since: 4,167,312; 4,167,313; 4,167,314.... He paints the number; he enunciates its numeral; he appropriates it by calling it by its name; and then he saves it for whoever studies his painting in the future. Only a close inspection of the bluish gloss of the painting reveals what goes into producing it. The most important aspect of the whole undertaking is this, however: Opalka knows—and so do we—that his project is bound to fail. Even if he found successors in his Kafkaesque obsession, even if all of mankind were to participate in keeping the counting going, the failure of the project is inevitable, the procession of numbers unstoppable. Nor is it helpful to object that counting toward infinity leaves reality behind as it reaches monstrous numbers that are completely detached from anything that is realistically countable. Most material perceptions make us just as aware of the trauma of the infinite as the process of counting. The most famous of these perceptions dates back to antiquity—to the third century BC and the man from Syracuse who was arguably the greatest mathematician that ever lived: Archimedes. The problem that Archimedes confronted was this: what is the ratio between the circumference of a circle and its diameter?

Some kind of answer to that question had already been implied in the Bible (1 Kings 7:23 and 2 Kings 4:2—both dating to the sixth century BC). However, this answer, *three*, which would make the circumference of a circle measure exactly three times its diameter, is obviously inaccurate. A regular hexagon inscribed *inside* a circle has three times the circumference of the circle's diameter because it can be subdivided into six equilateral triangles with the circle's radius as their side length—so the circumference of the circle *must* be longer than the threefold diameter.

Long before this, the ancient Egyptians came up with another answer. In the document known as the *Papyrus Ahmes* (dating from the seventeenth century BC), Egyptian scribes advise that multiplying a circle's diameter by 256/81 (just over 3.16) will give its circumference. What deliberations may have motivated the scholars of the time to opt for this value and what role number symbolism may have played in these deliberations is unknown. It may be noted, though, that 256 is the eighth power of 2, and 81 is the fourth power of 3.

However they arrived at their answer, the Egyptian scribes were also wrong. The first accurate calculation of the ratio between the circumference of a circle and its diameter remains one more feather in the cap of Archimedes—even if it does take the form of an *interminable calculation*.

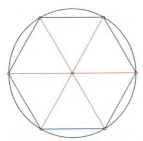

159 Circle and hexagon with three diameters that connect opposite corners of the hexagon.

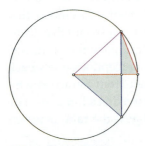

160 The long chord is bisected by the radius. The theorem of Pythagoras, applied to the two right triangles, enables us to calculate the length of the short chord from the length of the radius and that of the long chord.

161 Archimedes.

Two deliberations enabled Archimedes to reach his goal:[117] He already knew that the circumference of a regular hexagon inscribed in a circle was equal to three times that circle's diameter. He then succeeded in making an all-decisive second step: given the circumference of a regular polygon inscribed in a circle, he had a complex formula (bristling—to use the modern term—with radicals)[118] at his disposal that enabled him to calculate the circumference of a regular polygon with twice as many vertices inscribed in the same circle.

This second deliberation put Archimedes in a position to calculate the circumference of the dodecagon inscribed in a circle on the basis of the circumference of the hexagon inscribed in that circle. He then moved on to calculate the circumference of the polygon with 24 vertices inscribed in his circle. Next came that of the polygon with 48 vertices. He took this one step further, then broke off his calculation with the circumference of the polygon with 96 vertices inscribed in his original circle. He noted that this—in modern notation—amounted to the diameter multiplied by 3.14103… (the three dots indicate there are more decimals as yet uncalculated).

If Archimedes had had the computational resources at his disposal which are taken for granted today by every high-school student, he surely would have pursued his quarry beyond the 96-vertex polygon, to the polygons with 192, 384, 768 vertices, and so on. (Don't even try to work out what a polygon with 768 vertices would be in Greek!) However, he did not even have our modern decimal system to work with but could only use the cumbersome numerals of ancient Greece, which were expressed by letters of the alphabet. Given these additional hurdles, it is astonishing that Archimedes did not give up sooner than he did. He came to the conclusion that the circle's circumference is approximately 3.14 times its diameter—and to those who were not satisfied with this approximation, he let it be known that he had a method that enabled anybody to calculate the value to any degree of accuracy they desired.

One who took up the challenge many centuries later was Ludolph van Ceulen, a Dutch mathematical wizard who flourished around 1600. Van Ceulen started with a square inscribed in a circle and set about doubling the vertices with a missionary zeal that was as relentless as it was impressive. He ploughed on through the polygons with 8, 16, 32, 64 vertices all the way to a figure well in advance of four quintillion— 4,611,686,018,427,387,904 vertices. There he struck—he was an old man by now, having spent most of his life on his quest for the elusive ratio—which we have known as π since the seventeenth century—and concluded that the circumference of a circle is *approximately* 3.14159265358979323846264338327950288 times its diameter. The phenomenal accuracy of the ratio between circumference and diameter calculated to 36 significant figures did not obscure the fact that the accursed "approximately" could not be made to go away.

The ratio has been known as π since 1706, when the Welsh mathematician (and friend of Sir Isaac Newton) William Jones (1675–1749) coined the term from the initial letter of the Greek *perímetron*, meaning "boundary". In 1748, Swiss mathematician and physicist Leonhard Euler (1707–1783) emulated him, probably unwittingly, this time using the Greek word for circumference, *periféreia*.

With the help of the most powerful computers available today, π has been calculated to many more decimal places. At present, the record stands at some 1.24 trillion digits. If we assume there is room for 5,000 characters on one book page, π calculated to this degree of accuracy will fill 248 million pages, or almost a quarter of a million 1,000-page books. This would be a huge library in itself. On each of the 1,000 pages of each of these books, there would be a matter-of-fact string of decimal places of π. No trace of any sort of regularity in the sequence of numbers has been discovered even after the most exhaustive examination. On the contrary, the digits between 0 and 9 seem to be distributed absolutely at random just as the numbers 0 to 36 are at the casino in Monte Carlo.

What is most intriguing about such a determined effort at pinpointing π is that even the knowledge of a trillion decimal places reveals almost nothing at all about the absolute decimal realization of π. Continuing the process will not take us any further—even 100 quadrillion digits will not change this picture. An absolutely accurate value for π has still not been achieved, nor can it ever be achieved. Whatever extraordinary feats the computers of the future may achieve, they will never be able to calculate all the significant figures of π.

So what purpose does this enormous effort serve? There are two answers to this, one pragmatic and one philosophical.

The pragmatic answer is that computations such as this one are helpful in putting the latest developments in IT to the test. A machine that can calculate π to a quadrillion decimal places is putting its electrons through their paces—and nothing is as vital for computer manufacturers as that their troupe of electrons should adhere to a precise choreography.

The philosophical answer concerns the mysterious fact that the decimal realization of π is the way it is because the ratio of the circle's circumference to its diameter is the way it is: fraught with unanswerable riddles. The nature of both of these problems and the fact that both of them possess this same nature has a whiff of the absolute which is matched neither by the most minuscule elementary particles of the microcosm nor by the most distant galaxies of the macrocosm.

A comparison with *The Library of Babel* by Argentine author, critic and all-around man-of-letters Jorge Luis Borges (1899–1986) seems unavoidable here. Let us suppose that we were to go through the entire text of the library imagined above, replacing all the possible two-digit combinations in the decimal realization of π by a letter or punctuation mark—00 could be a space; 01, a; 02, b, etc. Our imaginary library, which houses the decimal realization of π as churned out by the most powerful and up-to-date computers, would have been converted into the *Library of All Conceivable Books* that Borges describes in his 1941 tale. The narrator, a librarian, describes the apparent chaos of his library, in which

> as is well known: for one reasonable line or one straightforward note there are leagues of insensate cacophony, of verbal farragoes and incoherencies.

However, Borges continues, this library is huge beyond all expectation; it is the *universe* and contains in its unfathomable depths

> all that can possibly be expressed, in every language there is, was or ever will be—all: the detailed history of the future, the autobiographies of the archangels, the faithful catalogue of the Library, thousands and thousands of false catalogues, the proof of the falsity of those catalogues, a proof of the falsity of the true catalogue, the Gnostic gospel of Basilides, the commentary upon that gospel, the commentary on the commentary of that gospel, the true story of your death, the translation of every book into every language, the interpolations of every book into all books, the treatise Bede could have written (but did not) on the mythology of the Saxon people, the lost books of Tacitus....

Yet this is only the beginning, for π is only one example[119] of the many quantities whose decimal realization tumbles into a bottomless abyss of digits following one upon another apparently at random. It is, moreover, a comparatively harmless example of the type, because behind its jumbled sequence of figures, a discernible law is at work. Metaphorically speaking, it is a Laplacean daemon that has been set the task of calculating the ratio between a circle's circumference and its diameter to an ever more exacting standard of accuracy. Law-based operations such as this are by no means all that may be concealed behind quantities that we strive to define ever more accurately as decimal numbers with ever more decimal places without ever making them absolute. For within this category, we can also find completely irregular quantities, aleatoric ones in the true sense of the word, whose decimal places follow upon one another in a completely and utterly random way. Brouwer found an apt metaphor when he referred to the formation of such quantities as the *free choices of a creative subject*. Their creation is governed by nothing but blind chance, quite literally *aleatoric* in that they do indeed come into being like the numbers of spots following one upon another as a gambler throws a die. It may very well be possible to arrive at an approximate decimal realization of a quantity generated in so utterly random a manner that stretches to ten, a hundred, one thousand, millions, arbitrarily many decimal places. Still, the result can never be anything other than an *approximation*. Even if the decimal places start off with an integer zero and if another thousand zeroes were to follow as next-in-line decimal places, an aleatoric quantity could never be relied on to settle for a simple and *absolute* zero. This still applies even if that quantity has a million zeroes as its first million decimal places, for it is still conceivable that in the billionth or the trillionth or the quadrillionth digit a one or a nine may crop up as a belated afterthought. It may even be the case that within this unending oscillation, within this irredeemable fluctuation of quantities, which ultimately depends on the fact that the process of counting is infinite, nestles the secret of life itself. After all, the most distinctive characteristic of life is spontaneity. Life is only possible against the backdrop of infinity.[120]

We get a distant view of the abyss of the infinite both in Opalka's artistic intuition and in his testimony. That abyss yawns at our very feet when we examine quantities with an unending stream of decimal places and experience the difference between *arbitrarily defined* standards of accuracy and *complete* accuracy.

The difference between arbitrarily defined standards of accuracy and complete accuracy in the specification of π may be insurmountable, but mathematicians have succeeded in the course of the development of their discipline to live with that difference. They have learned, quite literally, to reckon with it: how to add numbers to π and how to subtract them, how to multiply by π. They have even learned how to multiply π by itself—a remarkable achievement that bears little resemblance to multiplication as taught at school. It is not an operation confined to the multiplication table and a couple of additions.

This increasing mastery of calculating with values such as π—of which the abyss of the infinite is an inherent part—misled many mathematicians toward the end of the nineteenth century. In particular, it misled the German Richard Dedekind (1831–1916), who claimed that the borderline between operations involving quantities such as π and those involving numbers, such as 3, was no longer discernible. This being so, there was no justification in differentiating between simple numbers (meaning cardinal numbers derived from counting, 1, 2, 3, …) and quantities such as π whose decimal realization is powered by an interminable algorithm[121] or even the aleatoric quantities just discussed.

This view was vehemently criticized in the 1920s by the German-American Hermann Weyl (1885–1955), the last scholar who was able to survey the whole field of mathematics. The supporters of Dedekind were not just culpable of faulty logic, which might have been susceptible to correction, but had fallen into the trap of totally and perversely misconstruing and distorting the nature of these ultimately unfathomable quan-

162 Hermann Weyl.

tities. A mathematics that grounds itself on such warped premises was, according to Weyl, "built on sand". It was sheer presumption on the part of Dedekind to declare that the infinite could be subjected to calculation in the same way as numbers—such an assertion amounted to a denial of the nature of the infinite. Such utter recklessness must necessarily lead to the sacrifice of truthfulness, which turns the whole thing into a catastrophe not only in terms of the theory of cognition but, first and foremost, in terms of morality.[122]

An appropriate attitude toward infinity must, above all, be free from the delusion that it can be made the object of mathematical operations as if it were not radically different from numbers, as if it were a given quantity complete in itself like any other. In speaking about the infinite, we are up against a terminological border. It is ultimately beyond the reach of the human drive for knowledge—in the sense that we are, *pace* Dedekind and his illusions, unable to tame it. We still feel compelled to think about it, in Pascal's sense, and this is as it ought to be, because there lies concealed within it the *Library of Babel*, the universe.[123] There are few passages in world literature that describe the impact that infinity can have on our minds more graphically than the following key passage from the 1906 *bildungsroman* by the Austrian novelist Robert Musil (1880–1942), *The Confusions of Young Törless*. Wracked by inner turmoil, Törless lies down in a meadow at his boarding school, hoping for relief:

163 Robert Musil, age 17, upon leaving Eisenstadt Military College.

> The sky spread out above him, in that pale, ailing blue so typical of autumn, and little white round clouds scudded across it.
>
> Törless lay stretched out on his back and, dreaming vaguely, squinted between two treetops in front of him that were shedding their leaves. …
>
> And suddenly he noticed — and he felt as though this was happening for the first time — how high the sky really was.
>
> It came to him like a shock. Right above him there gleamed a little blue, unimaginably deep hole between the clouds.
>
> It seemed to him that if one had a long, long ladder, one should be able to climb into that hole. But the further he pushed his way in, lifting himself up with his gaze, the further away the blue, glowing background retreated. And yet he felt as though it should be possible to reach it and hold it, merely with one's gaze. The desire became painfully intense.
>
> It was as though the power of vision, strained to its limit, was flinging glances like arrows between the clouds, and as though, aim as far as they might, they always fell short.
>
> Törless thought about this now; he tried to remain as calm and sensible as he possibly could. "Of course there is no end." He kept his eyes fixed on the sky and said this out loud, as though to test the power of a magic spell. But without success; the words said nothing, or rather they said something quite different, as though they were referring to the same object, but to another strange, indifferent side of it.
>
> "Infinity!" Törless knew the word from maths class. He had never imagined anything particular by it. It was forever returning, someone must have invented it once, and since then it had become possible to calculate with it as surely as one did with something solid. It was whatever its value happened to be in the calculation; Törless had never ventured further than that.
>
> And now, all of a sudden, the idea flashed through him that there was something terribly unsettling about the word. It struck him as a concept that had formerly been tamed, one with which he had performed his daily little tricks, and which had now been suddenly unleashed. Something beyond understanding, something wild and

destructive seemed to have been put to sleep by the work of some clever inventor, and had now suddenly been woken to life, and grown terrible before him. There, in that sky, it now stood vividly above him and menaced and mocked.

Finally he closed his eyes, because the vision tormented him so.

(Translated from the German by Shaun Whiteside)

Notes

1   The account of these events given by the fifth century BC Greek historian Herodotus (Histories, 1:74) shows both armies panicked by the eclipse: both Lydians and Medes broke off their engagement when they saw this darkening of the day. They became more eager than they had been to conclude peace, and a reconciliation was brought about.

2   The Greek word *lógos*, for which Liddell and Scott's Greek-English Lexicon lists more than 100 distinct meanings, also signifies *ratio*, in particular the ratio between two numbers. It is in this sense the twin of the Latin word *ratio*.

3   There is nothing in the historical sources that would support this conjecture. On the contrary, the historical Pythagoras was almost certainly an *ontologist*, one who meditates on the principles of pure being, as opposed to a *hermeneutist*, one who is concerned with the problems of interpretation and understanding. However, as we know very little about either of them, we take the liberty of fashioning Pythagoras or Thales into ideal figures to serve our purposes.

4   It is said that charity begins at home. Perhaps we should look at communication with other species from a similar perspective. In the July 1, 2006, edition of *The Economist*, we find the article "Animal Behavior: A Stilted Story. Ants Find Their Way Home Using Pedometers", a report on a paper in that week's *Science*. The article concerned describes an experiment dealing with *Cataglyphis*, a species of desert ant. This experiment suggests that *Cataglyphis*, which navigates by the sun, somehow manages to count its steps on its outward journey and to tally that figure on its return, thus allowing it an instant and accurate answer to the query "Are we there yet?" The journal concludes by saying, "Ants may not be very bright but it seems they have a head for figures". Maybe they do, and maybe they are not alone. How many more of our cousins in our very own animal kingdom have a head for figures? I am not aware that any systematic study of number-related behavior among non-human animals has as yet been undertaken.

5   The notion that even numbers are "evil" and odd ones "good" plays a decisive role in the following multiplication technique, for which only the mastery of adding, halving and doubling is required. Assuming that we want to multiply the numbers 75 and 57, we proceed as follows:

75	57
37	114
~~18~~	~~228~~
9	456
~~4~~	~~912~~
~~2~~	~~1824~~
1	3648

In the column on the left, the first factor (75) is continually halved (remainders are consistently ignored); in the one on the right, the second factor (57) is continually doubled. Recalling that "even numbers are evil", we delete all lines containing an even number on the left. Then we add up all the remaining figures in the right column that have not been deleted. The addition 57 + 114 + 456 + 3,648 yields 4,275 as the product for the multiplication of 75 · 57, which is indeed the figure we were expecting to obtain.

For unbelievers, we will reverse the procedure by swapping round the factors:

57	75
28	150
14	300
7	600
3	1200
1	2400

Lines two and three have to be deleted, because the even numbers 28 and 14 are "evil", and the addition of 75 + 600 + 1,200 + 2,400 again yields the same product as above.

It goes without saying that we can point out why this technique always yields the correct result, without having to accept the notion of "good" or "evil" numbers. The fact remains that generations of reckoners based their multiplications on this notion, which does have an aura of magic about it. In view of how blindly we accept the results of computer operations, we should indeed beware of labeling as backward those who successfully manipulated numbers even if they did believe that odd ones were good and even ones evil.

Even enlightened thinkers, such as Leibniz, found a formula seemingly proving that odd numbers were more pleasing in the sight of God. If the reciprocal values of odd numbers—i.e., 1/1, 1/3, 1/5, 1/7, 1/9, 1/11—are alternately subtracted and added, the "infinite sum" approaches a quarter of the value of $\pi = 3.14159\ldots$. It goes without saying that it is actually impossible to calculate this infinite sum. One must make do with finite subtotals, and consequently all one ever gets is approximate values. With reference to the infinite sum discussed above, these approximate values converge to $\pi/4$ pitifully slowly, as Leibniz' adversary Newton dismissively remarked.

6   Matthew tells of a similar miracle where Jesus fed five thousand men (to say nothing of the women and children) with five loaves plus two fish (again a total of seven objects). On this occasion the scraps filled twelve baskets. (Mt 14, 17-21).

7   For the significance of the number forty, I am indebted to Anna Bergmann's *Der entseelte Patient. Die moderne Medizin und der Tod* (Berlin, 2004), in which she traces the lineage of the concentration camp back to quarantine management from 1374 on.

8   The Pythagoreans believed that the divisors of a number enabled them to determine the *quality* of that number and of everything that number represented. Some numbers, such as twelve (with its divisors 1, 2, 3, 4, 6 and 12) or thirty (with the divisors 1, 2, 3, 5, 10, 15 and 30), are "superabundant"—that

is, they "overflow" in a sense because the sum of their divisors (excluding the number itself) is larger than they are. In the case of twelve, the sum of 1, 2, 3, 4 and 6 is 16; in the case of thirty, the sum of 1, 2, 3, 5, 10 and 15 is 36.

Other numbers, such as ten (with its divisors 1, 2, 5 and 10) or sixteen (divisors 1, 2, 4, 8 and 16), are deficient—that is, they are "thirsty" because the sum of their divisors (excluding the number itself) is smaller than they are. In the case of ten, $1 + 2 + 5 = 8$; for sixteen, the sum of $1 + 2 + 4 + 8$ is 15.

The Pythagoreans regarded a number as "perfect" if the sum of its divisors (excluding the number itself) is identical with that number. Six, for instance, with the divisors (1, 2, 3 and 6) and 28 (divisors 1, 2, 4, 7, 14 and 28) are perfect, because $1 + 2 + 3 = 6$ and $1 + 2 + 4 + 7 + 14 = 28$. The next perfect numbers are 496; 8128; 33,550,336 and 8,589,869,056. It is impossible to calculate how many there are or whether there may be odd ones among them.

"Friendly" or "amicable" was an epithet given by Pythagoras to certain pairs of numbers; for it to be awarded, the sum of divisors of one member of the pair (again excluding the number itself) has to yield the value of the other member—and vice versa. Such a pair of friendly numbers known to the Pythagoreans was 220 (with the divisors 1, 2, 4, 5, 10, 11, 20, 22, 44, 55, 110 and 220) and 284 (divisors 1, 2, 4, 71, 142 and 284). These sums are $1 + 2 + 4 + 5 + 10 + 11 + 20 + 22 + 44 + 55 + 110 = 284$ and $1 + 2 + 4 + 71 + 142 = 220$.

Friendly numbers were symbolic of souls befriending one another. This idea outlived antiquity. A medieval number mystic, for example, recommended writing the numbers 220 and 284 on slips of paper, making sure that the beloved ingests the slip with the smaller number, and eating the one with the larger number oneself. The proponent of this prescription claims to have first-hand experience of its erotic effects.

9   Euclid conducted the proof as follows: if someone presented him with a list of a finite number of prime numbers, Euclid asked them to multiply all the prime numbers by one another and to add one to the product. The number obtained in this way cannot be divisible by any of the prime numbers in the list, because there will always be a remainder of one from divisions by any of them. The number obtained in this way is either divisible by a prime number not contained in the list or is itself a prime number.

The two examples $2 \cdot 3 \cdot 5 + 1 = 31$ and $2 \cdot 3 \cdot 5 \cdot 7 + 1 = 211$ may easily tempt one to conclude that the number generated by Euclid is itself a prime number. But, as can be seen from the example $2 \cdot 3 \cdot 5 \cdot 7 \cdot 13 + 1 = 30,031$, this is not always the case. True, Euclid has made sure that 30,031 is neither divisible by two nor by three, and the same applies to five, seven and thirteen. Yet it is far from being a prime number, being in fact the product of the prime numbers 59 and 509.

10   There are 880 different magic squares with four columns and four rows. For magic squares with five columns and five rows, the number of different realizations exceeds ten million.

11   For instance, the letters that follow after θ (theta) = 9, ι (iota), κ (kappa), λ (lambda), μ (mu) denote 10, 20, 30, 40, respectively. The combination $\lambda\beta$ (lambda beta) = 32, whereas $\beta\lambda$ (beta lambda) = 23. The letter π, reserved since Euler's days to 3.14159…, was understood by Archimedes to mean 80.

12   The first acrostics or acronyms are to be found in Babylonian prayers; they certainly served a magic function.

13   It is obvious that there is a direct connection between those seven words on the one hand and the six days of creation and a day's rest on the other.

14   If we had stopped short one step earlier in each case rather than at a heptatonic scale, we would have arrived at the pentatonic one of the five tones D, E, G, A, C—the tonal system of quite a few non-European civilizations. If one substitutes A♭ for D, one gets the pentatonic scale of A♭, B♭, D♭, E♭, G♭, which corresponds to the black keys of the piano. Chopin's famous "Black Keys" Etude in G♭ major, Op. 10 No. 5, is inspired by the pentatonic scale.

15   The hope for "redemption" from the Pythagorean comma is bound to be frustrated. Any attempt to arrive at the same note with the help of ascending fifths (or descending fourths) is doomed to failure. The intervals formed by fundamentals with their ascending fifths are characterized by the fact that the numerator of the fraction in question contains a number that is a power of three, therefore an odd number, whereas their denominator contains a power of two, therefore an even number. For the intervals formed by fundamentals with their descending fourths, the reverse is true. Since an even number can never coincide with an odd one, the Pythagorean comma is ineradicable, no matter how far the ascent of the fifths (or the descent of the fourths) is pursued.

16   The difference between the natural seventh and the minor seventh mentioned earlier is computed in a manner similar to the one employed for the Pythagorean or the syntonic commas:

$$\frac{9}{5} \cdot \frac{4}{7} = \frac{36}{35} = 1.0286.$$

Therefore it amounts to roughly 2.9%.

17   Resigning oneself to only two dimensions is reminiscent—to entertain for a moment an idea that may be no more than a superficial analogy—of the string theory of modern theoretical physics; in it the universe has ten or more dimensions, most of which are so curved in on themselves that we cannot perceive them.

18   Strictly speaking, we have to differentiate between two possibilities: our sense of hearing identifies an interval for what it is but marks it as "out of tune". What we have in mind at present is the second case: our sense of hearing labels intervals that are slightly out of tune as "pure".

19   In the perception of the underlying facts, Euler was preceded by Conrad Henfling (1648–1716), a theoretician of music, who corresponded with Leibniz on the subject.

20   It is this definition that enables us to work out the decimal digits of the twelfth root of 2. Let us assume that we already know that the product of 1.059 multiplied with itself 12 times is smaller than 2, whereas the same operation performed with 1.060 yields a product bigger than 2. We therefore accept 1.059 as the closest approximation to the twelfth root of 2 calculated to three digits after the decimal point. In the next step we perform the same operation with

1.0590, 1.0591, 1.0592, 1.0593, ..., 1.0599. We realize that 1.0594 multiplied with itself 12 times yields a product smaller than 2, whereas the same operation performed with 1.0595 yields a product bigger than 2. Therefore, we accept 1.0594 as the closest approximation to the twelfth root of 2 calculated to four decimal digits. This operation can obviously be repeated ad lib, which means that the twelfth root of 2 can be calculated to any arbitrarily defined cut-off point.

21 If one maps the Eulerian tone lattice on to this torus, the effects of the Pythagorean comma and of the dieses will cause the lattice nodes—which represent the tones—to spread quickly, evenly and densely. According to the *Theory of Equal Distribution*, which was first formulated by the most distinguished mathematician of the twentieth century, Hermann Weyl, and developed by the Austrian mathematician Edmund Hlawka, the tones of the Eulerian lattice are "equally distributed" on the torus of equal-tempered intonation even "modulo octaves".

22 This is the term commonly applied to this type of spatial curvature in Walter Wunderlich's Vienna School of Descriptive Geometry. In Hermann Weyl's paper on equal distribution, the term *rotoid* is replaced by the cartographical term *loxodrome* of the torus, literally an "oblique runner", because it intersects the meridian circles of the torus on which it is positioned at a predetermined angle.

23 Starting from F♯, one ends up, strictly speaking, with D♯♯♯♯.

24 Admittedly, the case we are trying to make is not entirely watertight. It is based on the assumption that our ear listens to the fugue theme almost as it would to a Schönbergian 12-tone row, i.e., by processing interval upon interval. We would claim that our tendency to listen in this way is reinforced by two facts:

   • Pianists cannot play their instrument in such a way as to modify their tones to reflect underlying harmonies. This in its turn causes us to refrain from seeking these harmonies.

   • In a rare instance of the composer choosing to curtail interpretive freedom, Bach prescribed a tempo for this fugue, "Largo". By doing so, he encourages us to follow Johann Matheson's precepts and to concentrate on intervals.

   If one were to play the fugue theme on a string instrument such as a violin, one would be more likely to return to the original F♯.

25 This is immediately followed by a dialogue with the youth Octavian, in which the Marschallin returns twice to the mysterious nature of Time:

   It seems to me
   That I am made to feel the frailty of the temporal world
   In my heart of hearts:
   All I cannot do—never to cling,
   Never to catch and hold.
   And everything runs through my hands,
   Melts from my touch,

Dissolves like mist and dreams.
. . . .
Time is very strange.
We live our lives, time is nothing.
Then all at once
It occupies all our senses:
Surrounds us, is inside us,
Falls softly on our faces, on our mirrors,
Turns my temples grey.
And between you and me it falls,
Noiselessly, as in an hourglass.

26 In strict analogy to the flying arrow, Zeno argues the case of the notorious Achilles and the tortoise paradox. He claims, as is well known, that if the slowest of all creatures, the tortoise, were given a head start in a race against the fastest known runner, Achilles, it was impossible for him to overtake it, regardless of the competitors' differing athletic competence. The pursuer always needs to first reach the point the pursued has just vacated. Mathematicians are prone to believe that a reference to a geometric series will make the paradox dissolve into thin air. They overlook the fact that the computational sleight of hand does not in any way affect the logical aporia which culminates in the question: what logical fallacy enabled Zeno to devise this nonsense in which reason—if we accept his premises—gets hopelessly trapped?

27 The sun actually takes a little longer than a stellar day between these apexes because the earth's orbiting around the sun results in an apparent "delay" of approximately four minutes. The precise figure is obtained if we divide the number of minutes of one day, $24 \cdot 60 = 1,440$, by 365, the days of one year.

28 In antiquity, *New Moon* referred more fittingly to the phase when the ghost of a sickle ascends the heavens on the first night of its renewed visibility. The term *New Moon* should be replaced by *Black Moon*—but the prospects of such a change catching on are practically nil.

29 The *synodic* month lasts more than two days longer than the *sidereal* one—the time the moon actually takes to complete its orbit round the earth. The difference between synodic and sidereal lengths is due to the orbital movement of the earth-moon system around the sun. Because the earth drags the moon with it as it orbits around the sun, the end of each sidereal month sees a change in the angle at which the sun lights up the moon. The orbital plane of the moon around the earth intersects the orbital plane of the earth around the sun at an angle. Although this angle is very shallow, it still contributes to the rarity of lunar and solar eclipses. These occur at oppositions and conjunctions of moon and sun but only when the line of intersection of the two orbital planes coincides precisely with the earth-sun line. The line of intersection rotates around the earth in a rhythm of 27 days and 5 hours. It was again Babylonian astronomers who were the first to observe that solar and lunar eclipses were regular occurrences, displaying a cycle of 223 synodic months or of 18 years and 10 or 11 days (depending on whether that period included four or five leap years). This is known as the *saros* cycle and owes its length to the rough match between the following multiplications:

$$223 \cdot (29 + \tfrac{1}{2}) = 6578 + \tfrac{1}{2} \text{ and } 242 \cdot (27 + 5/24) = 6584 + 5/12.$$

Saros was known not only to Thales of Miletus but also to the Mayans, who based calendars of great exactness on it.

30 This is the date in spring when the sun passes the First Point in Aries. Because of precession (a slight wobbling of the earth's axis that is only observable relative to the stable sun-earth system), the First Point moves full circle along the Zodiac in the course of 26,000 years and is now—a couple of centuries after its original fixation in Aries—in the constellation of Pisces.

31 The Jewish calendar, which is still significant today for the fixation of Jewish holidays and religious feasts, takes the Babylonian one as its model. The 3rd, 6th, 8th, 11th, 14th, 17th and 19th year of the 19-year cycle are augmented by the leap month, and the number of days in each year varies according to a complicated system between 353, 354 and 355 in ordinary years; 383, 384 and 385 days in leap years.

32 Because $365 \cdot 4 = 1,460$, the Egyptian year wanders through the four seasons in a period of 1,460 years, the so-called *Sothis* period. Sothis, the Greek name for Sirius, is a word loaned from Egyptian, where *Sepdet* signifies "female herald of the Nile".

33 The intercalated day was not February 30 but February 24, the day after the Roman feast of Terminalia on February 23, originally the last day of the year. The Roman calendar before Caesar was in a dreadful mess; it was based on the lunar year, and in leap years, a leap month, Mercedonius, was intercalated after the Terminalia.

34 Guardianship of the calendar had passed from the pontifices of ancient Rome to the Church, which had the side effect—welcome, from the Church's point of view—of giving the Church the opportunity to demonstrate its sovereignty over time. The most challenging part of the task was to calculate the date of Easter in any given year, past or future. As Easter has to be celebrated on the first Sunday following the first Full Moon after the spring equinox (this has to do with a combination of Sunday as the day of the Resurrection, the date of the Jewish Passover festival and the lunar calendar), both the days of the week of any given year and the phases of the moon need to be taken into account when calculating the date of Easter. For the Julian calendar, this results in a rhythm of $28 \cdot 19 = 532$ years, after which the sequence of calendar days for Easter Sunday repeats itself.

35 As far as the division into months was concerned, the French revolutionary calendar was modeled on the Egyptian one.

36 Vendémiaire, "vintage", from Latin *vindemia*: September 22 to October 21; Brumaire "misty", from French *brume*: October 22 to November 20; Frimaire, "frosty", from French *frimas*: November 21 to December 20; Nivôse, "snowy", from Latin *nivosus*: December 21 to January 19; Pluviôse, "rainy", from Latin *pluviosus*: January 20 to February 18; Ventôse, "windy", from Latin *ventosus*: February 19 to March 20; Germinal, "seedtime", from Latin *germen*: March 21 to April 19; Floréal, "blossom", from Latin *flos*: April 20 to May 19; Prairial, "meadow", from French *prairie*: May 20 to June 18; Messidor, "harvest", from

Latin *messis* + (presumably) Greek *doron*, "gift": June 19 to July 18; Thermidor, "hot", from Greek *thermós*: July 19 to August 17; Fructidor, "fruits", from Latin *fructus*: August 18 to September 16.

37 Even in the great reform of Gregory XIII, Thursday October 4 was followed by Friday October 15, 1582.

38 For his mechanical experiments, Galileo Galilei used a kind of water clock as a timekeeping device. Galileo would call out to his assistant to signal the beginning of an experiment; the assistant thereupon caused water to flow regularly into a receptacle until the end of the experiment. The time that elapsed between these two points in time was then literally weighed.

39 The *pontifex maximus* Julius Caesar has been joined by his successors, the Roman pontiffs, in his abdication as Guardian of Time. This role has been assigned to physicists for the time being. For the layman the difference is presumably negligible. In antiquity, the ritual proceedings of members of the priest caste, shrouded in clouds of incense, were experienced as numinous and enigmatic by the common people. Likewise, the experts' mumbo-jumbo that issues from today's temples, the hi-tech labs, is intelligible only to those who have been initiated into the pertinent jargonistic formulae. We might consider the following definition of a second as an example: "The second (symbol s) is a unit of time, and one of seven SI base units. It is defined as the duration of 9,192,631,770 periods of the radiation corresponding to the transition between the two hyperfine levels of the ground state of the caesium-133 atom at zero degrees Kelvin." Fairly trips off the tongue, doesn't it?

40 It is fairly certain that AD 1 has nothing to do with Jesus' actual year of birth. In AD 1, the six- or seven-year-old Jesus was presumably already busy studying the Torah.

41 It is beside the point to try and discredit the hypothetical dating of Creation according to the Jewish chronology by pointing to the existence of obviously much older fossils. The rhythm of counting is freely adaptable. The retrograde counting of years drifts loose from the mundane consideration of earth's revolutions around the sun (which, in any case, can only be measured on an ad-hoc basis), and its speed therefore varies tremendously from civilization to civilization.

42 In calculating the area of a circle, the Egyptians made do with a reasonably accurate approximation technique, which they however believed to be exact. Our source for this is the *Rhind Mathematical Papyrus*, named after A. Henry Rhind, the Scottish Egyptologist, who purchased it in Luxor in 1858. It is also known as the *Ahmes Papyrus* after the scribe who copied it in about 1650 BC. The first section bears the heading "Correct method of reckoning, for grasping the meaning of things and knowing everything that is, obscurities and all secrets." In Task 50, a rule is given to calculate the area of a circle: "Cut off 1/9 of the circle's diameter and construct a square on the remainder. The area of this square will be identical with that of the circle."

43 "Centimeter" was, of course, unknown as a measurement of length at the time. We use it here for simplicity's sake, and any type of unit will do equally well in our context.

44   Of course, the numbers for the sides of the triangle have been arbitrarily determined. Apart from this fact, this is the simplest proof of the Pythagorean theorem.

45   It is, however, possible to check these numbers out by means of Bhaskara's figure. The numbers relevant for the area of the inner square are

$$(13{,}500 - 12{,}709) \cdot (13{,}500 - 12{,}709) = 791 \cdot 791 = 625{,}681.$$

Those for the four triangles in question are

$$2 \cdot (13{,}500 \cdot 12{,}709) = 343{,}143{,}000.$$

If we add these two numbers, $625{,}681 + 343{,}143{,}000 = 343{,}768{,}681$, we get the area of the outer square, the one with the third side of the triangle as its side length. As $18{,}541 \cdot 18{,}541$ indeed equals $343{,}768{,}681$, the triangle with sides as specified above is indeed a rectangular one.

46   The trick consists in starting off with two non-identical numbers chosen at random, e.g., 125 and 54. The difference of their squares, in our example $(125 \cdot 125) - (54 \cdot 54) = 12{,}709$, and their doubled product, in our example $2 \cdot (125 \cdot 54) = 13{,}500$, are suitable numbers for the cathetes or legs of the triangle. The length of the hypotenuse will be the numerical value of the sum of the two squared original numbers, in our example $(125 \cdot 125) + (54 \cdot 54) = 18{,}541$.

47   An obvious question is why the right angle is apparently instrumental in bringing abstract numbers into close contact with descriptive figures. It may be helpful in this context to consider some aspects of the basic arithmetical operations of addition, subtraction and multiplication.

Let us start with an addition of two addends: 7 and 9. This operation is easy to visualize as the elongation of a line measuring 7 units by another 9 units to produce a line measuring 16 units. Similarly, for the subtraction of 7 from 9, all we need to do is plot 9 units and 7 units from one and the same point on one line; the remaining two units represent the difference between 9 and 7. However, multiplication is not as straightforward. The multiplication of $7 \cdot 9$ is not simply shorthand for a sevenfold addition of 9; it has, at the same time, the descriptive force of the area of a rectangle. A line one centimeter long, a standard line, also defines a square whose sides are one centimeter long, the so-called unit square. Whether we use unit lines or unit squares for the addition or subtraction of 7 and 9 makes no difference whatever. Yet the multiplication of $7 \cdot 9$ can be visualized by arranging the result, 63, in the form of a rectangle 7 centimeters long and 9 centimeters wide. It represents an area of $7 \cdot 9 = 63$ square centimeters.

Why, we might ask, do we confine ourselves to using squares constructed on the base of the standard line as area units? Why not use a lozenge-shaped parallelogram with four standard lines as sides and no interior right angle? Why not define this shape as area unit? The *standard rhombus*? In that case we would be calculating the areas of parallelograms instead of rectangles in working out the product. The parallelograms would have the same pairs of interior angles as the standard lozenge. Multiplying the lengths of the two sides of the parallelogram with each other would tell us how many standard lozenges fit into the parallelogram—we would have worked out the area in *rhombus centimeters*.

$$7 \cdot 9 = 63 \qquad\qquad 7 \cdot 9 = 63$$

However, as opposed to rectangles, the above parallelograms have diagonals of different lengths. One diagonal would produce a pair of congruent triangles completely different from the result of the other diagonal.

In this difference between a non-rectangular parallelogram and a rectangle resides the mysterious uniqueness of the right angle. In the case of the parallelogram, to arrive at a figure reminiscent of Bhaskara's construction we need both the pair of triangles generated by cutting along the shorter diagonal and the one generated by cutting along the longer diagonal. In this figure, the two pairs of diagonals form the perimeter of a large parallelogram. True, at its center there is a lozenge with sides the length of the difference between the sides of the large parallelogram and with the interior angles of the standard rhombus precisely in the manner in which a small, slightly rotated square appeared in Bhaskara's figure. Yet the large surrounding parallelogram has interior angles completely different from those of the standard rhombus.

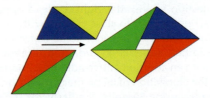

It would therefore be a mistake to multiply the lengths of the diagonals with a view to obtaining in this way the area of the large parallelogram as a multiple of the standard rhombus, that is to say, in *rhombus centimeters.*

Bhaskara's demonstration, which bridges the gulf between numbers and figures, can be successful only if the area is defined in unit squares, all of whose interior angles are right ones.

48   Leonardo of Pisa, the Renaissance scholar, recasts the problem in the martial mold of a spear leaning against a tower.

49   The degrees of an angle with their symbol of a small superscript circle are another innovation for which we are indebted to the Babylonians. They subdivided the circle into 360 equally long arc segments and expressed the size of an angle through the number of segments it encloses when its vertex is superimposed upon the center of the circle. As a quadrant encloses $360/4 = 90$ of these segments, a right angle is $90°$.

50   If in any non-specific triangle, we draw a line from a vertex in such a manner that it encloses a right angle with the opposing side, it is obvious by the same reasoning that the sum of a triangle's internal angles is identical with the sum of two right angles, i.e., $180°$.

51   The word *sine* has come into being in a curiously circuitous way. Indian mathematicians called the ratio between the half chord of an arc and the circle's

radius the *ardhajiwa* (literally, *half chord*) of that angle whose doubling opens up the complete arc. The Arabs took over the foreign word and mangled it to *djib*. As Arabic script dispenses with vowels, the word was indistinguishable in its written form from *djaib* = *bosom*. When Arabic mathematical textbooks were translated into Latin, the latter proved the more attractive option for translators and consequently they sought to render the word with *sinus* = bosom, curve etc.

52 Ptolemy used two methods for his purpose. He knew by how much an opposing side was shortened when the angle was halved, and he also knew how to work out the length of an opposing side when two angles were superimposed on one another provided he knew the lengths of the opposing side of each individual angle before the superimposition. Both methods follow from Pythagoras' theorem—that is to say, from Bhaskara's figure.

53 The unit of length measurement in antiquity was the *stadion*, like the mile a rather elusive character. We follow modern usage here and have also rounded numbers to make the underlying principles of calculation as clear as possible. In actual fact, Eratosthenes, in his geophysical survey, was off by 12%: an error attributable solely to the fact that the distance Syene–Alexandria had not been accurately measured.

54 The cartographic method outlined here is called the *Gnomonic* (from *gnomon*, the vertical staff whose shadow indicates time on sundials) or *Gnomic* or *Central* projection and may have been developed by Thales. One among a variety of projections, it has the drawbacks that distortion increases rapidly away from the (distortion-free) center and that it is capable only of showing less than one hemisphere. It has advantages for navigators (and aviators) in that it indicates the shortest connections between two points by showing great circle-paths as straight lines.

55 Eratosthenes' measurements are not in themselves sufficient to prove conclusively the earth's spherical shape. The shadow at Alexandria might after all be the result of a hypothetical proximity of sun and earth, which would make the sun's rays fan out relative to the earth.

56 Similar differences between the calculation of distances in space and their concurrent measurement have been observed in the vicinity of the sun and even more so near the ominous black holes. These differences corroborate Einstein's postulate that great masses are surrounded by warped space.

57 The distance from the earth to the moon is, approximately speaking, 60 times the radius of the earth—this being a mean value because the moon revolves around the earth on an elliptical, not a circular, orbit.

58 Hipparchus knew that the shadow cast by the earth on the moon had to be cone-shaped—rather than cylindrical—because the diameter of that shadow shrinks with the earth's distance until it is point-size. The internal angle of that cone—or, put another way, the rate at which the shadow's diameter shrinks with distance—can be deduced from the ostensible size of the sun's disk. Seen from the earth, the diameter of the sun's disk is almost exactly one half of a degree, and this corresponds almost exactly to the internal angle of the cone of the earth's shadow. Hipparchus, moreover, computed from the curvature of

the earth's shadow on the moon how much smaller the moon's radius had to be relative to that of the earth. The moon's radius is in fact roughly a quarter. As Eratosthenes had already established the earth's radius, all that Hipparchus needed to do was to calculate how far away a sphere with a radius a quarter that of earth's had to be for its disk to appear the size with which the moon presents itself to us. In this way, he calculated the distance between the earth and the moon with admirable accuracy.

59   The dates of Venus transits subsequent to the dates mentioned earlier are as follows: December 9, 1874, and December 6, 1882; June 8, 2004, and June 6, 2012; December 11, 2117, and December 8, 2125.

60   The unit Bessel used to indicate the distance was the so-called astronomical unit, the equivalent to earth's median distance from the sun, i.e. roughly 93.3 million miles.

61   For the measurement of distances of this order, physical methods take over from geometrical ones. One method is to calculate the brightness of a star on the basis of its color and compare this to the brightness of a star such as 61 Cygni. If it is perceived to have a lesser radiance than 61 Cygni's luminescence despite having a greater brightness, it must be farther away from earth than 61 Cygni in proportion to these quantities.

62   In earthly terms, this amounts to a distance of a monstrous 15,000,000,000, 000,000,000 (15 trillion) miles.

63   The factor of the slowdown equals the sine of the geographical latitude of the pendulum's location. If we were to suspend Foucault's pendulum at the equator, there would be no apparent rotation of the plane of oscillation because the whole pendulum would revolve in tandem with the earth.

64   Another project, as alien as it could well be in spirit from Camus' heroics but also designed to overcome our cosmic isolation is advanced by scientists with an almost religious zeal for the development of artificial intelligence. Their goal is to ferry computers, because they are less dependent on terrestrial conditions than human beings, into outer space, perhaps in a not too distant future and, using earth as their base, proceed to colonize the cosmos. Human beings made of flesh and blood no longer form an integral part of the project. At best they will be kept as mascots for a time to be ultimately subjected to mass extermination as *faulty prototypes*. The proponents of such appalling scenarios, some of them highly respected professors at prestigious places of higher learning, do not seem to be bothered by the argument that they are taking over where Hitler was forced to bow out ignominiously. They are in fact expanding the scope of Hitler's original undertaking to include all of humankind. One is certainly reminded of those scientists in the Third Reich who advocated the *extermination of races unworthy of life (lebensunwert)*. It is chilling to remember that one of these, Konrad Lorenz, even went on to win a Nobel Prize.

65   A straight line is drawn at a right angle to the line connecting the internal point and the center of the circle that intersects with the circle at the tangential points. The point of intersection between the line connecting the center of the circle with the internal point and the tangents will be the original external point.

66   For a bibliography on the hollow-world issue, see Ruth S. Freitag: *Hollow Earth Theories*, http://www.loc.gov/rr/scitech/SciRefGuides/hollowearth.html

67   The circle does not necessarily represent a cross section of the globe. With a touch of playfulness, we could equally well make it out to be the cross section of the human observer's head (idealized to the shape of a circle for simplicity's sake). The whole drama is then enacted on this observer's inner stage.

68   The modern, highly controversial *String Theory* presupposes a universe with many more than three dimensions to accommodate the great number of theoretical constructs.

69   With this interpretation Descartes appeared to give moral sanction to the most revolting animal experiments.

70   The origins of the Roman numerals are rather confusing. We may assume with a high degree of probability that the numerals were originally modeled on letters of the Greek alphabet. The vertical stroke of "I" indicates the unit; "V" and "X" are symbols of bundles. For C and M to have assumed their shapes, the fact that they thus represent the first letters of *centum* (= hundred) and *mille* (= thousand) was presumably important. In early Roman documents we find the Greek letter Φ symbolizing 1000, which explains the use of D for one half of 1000.

71   The Romans developed a specifically Roman form of the abacus for the use of merchants, engineers, tax collectors and so on, which represented a considerable advance on traditional Greek counting boards and predated the Chinese *Suan Pan* by several centuries. For details of the reconstructed specimen at the London Science Museum, see Wikipedia's *Roman Abacus*.

72   Bearing in mind how we reached the equation 75 (base ten) = 1001011_2 (base two) and the fact that what we are doing in multiplying 75 by 57 is in effect adding up individual products ($57 \cdot 1_2 = 57 \cdot 1 = 57$), ($57 \cdot 10_2 = 57 \cdot 2 = 114$), ($57 \cdot 1000_2 = 57 \cdot 8 = 456$), ($57 \cdot 1000000_2 = 57 \cdot 64 = 3{,}648$), we may be reminded of the "numerical-symbolist" rules for multiplying by halving and doubling with the *magic formula*, according to which even numbers are "evil", in Chapter 1.

73   A place value system that is as simple as binary, yet more attractive in many ways, has been overlooked by Leibniz and many of his successors: a triadic system with *three* as its base. The opportunity at least to provide an outline of this highly elegant place value system here is too good to be missed.

        The triadic representation of the number 57 is obtained along lines similar to those followed for binary but will involve three ciphers: o, which will again represent zero, p, which in the system under discussion represents 1 and q, which in this system signals the absence of 1 (mathematically speaking, minus 1).

        First we note that 57 is divisible by 3. The units digit of the triadic representation of 57 is therefore o. We then divide ($57 - o = 57$) by 3 and obtain the number 19. Because 19 is 18 + 1, it is divisible by 3 after we subtract 1. Therefore, the units digit of 19 is p. We divide ($19 - p = 18$) by 3 and obtain the number 6. Because 6 is divisible by 3, the units digit of its triadic representation is o. We divide ($6 - o = 6$) by 3 and obtain the number 2. Because 2, in order to be divisible by 3, needs to be augmented by 1, the units digit of 2 is q (because it is 1 short of a number divisible by 3). We divide ($2 + q = 3$) by 3 and obtain the

number 1, which is identical with the cipher p of the triadic system. All five
steps are summarized in the table below:

$$57 = 57 + o$$

$57 = 19 \cdot 3$	$19 = 18 + p$
$18 = 6 \cdot 3$	$6 = 6 + o$
$6 = 2 \cdot 3$	$2 = 3 + q$
$3 = 1 \cdot 3$	$1 = p$

The triadic representation of 57 is pqopo.

Conversely, a number in triadic representation, e.g., poqpo can be instantly
converted into decadic by multiplying the ciphers with the power of 3 that
corresponds to their place value and then forming the sum:

$$poqpo = o + p \cdot 3 + q \cdot 9 + o \cdot 27 + p \cdot 81$$
$$= 0 + 1 \cdot 3 - 1 \cdot 9 + 0 \cdot 27 + 1 \cdot 81$$
$$= 3 - 9 + 81 = 75.$$

In electronic terms, the triadic system makes perfect sense. In addition to
the uncharged or unpolarized state, which corresponds to the cipher o, we
distinguish between two more states—the positively charged or northbound
state, which corresponds to the cipher p, and the negatively charged or south-
bound state, which corresponds to the cipher q.

One of the advantages of the triadic system over binary is that digits do not
multiply so quickly as numbers get bigger. For the first 12 numbers, this is not
yet clearly in evidence. Beginning with zero, this is what 0 to 11 would look
like in the binary system:

0, 1, 10, 11, 100, 101, 110, 111, 1000, 1001, 1010, 1011.

Their triadic representations are only a trifle more economical:

o, p, pq, po, pp, pqq, pqo, pqp, poq, poo, pop, ppq.

However, with really big numbers such as a billion (in base ten notation
ten digits: 1,000,000,000) the difference becomes noticeable indeed. In binary,
1,000,000,000 = 111011100110101100101000000000, a 30-digit number, whereas
in the triadic system, 1,000,000,000 = poqqpqopqoopqpopoooop, a mere 20
digits.

The balance tilts the more in favor of the triadic system, the more numbers
increase in size. It is, of course, very big numbers that form the typical material that computers process.

Calculation rules for addition and multiplication are as simple in the triadic system as they are in the binary one. The following rules of "one-plus-
one" will make it possible to carry out each and any addition:

$p + p = pq,$	$p + o = p,$	$p + q = o,$
$o + p = p,$	$o + o = o,$	$o + q = q,$
$q + p = o,$	$q + o = q,$	$q + q = qp.$

To add up the two numbers 75 = poqpo and 57 = pqopo, the two numbers
are written one below the other with their place values in vertical alignment.
Starting with the units digit, we form the sums of the different digits; in the

case of p + p = pq and q + q = qp, we write down the singles digit q (or p, as the case may be) and "carry" the threes digit to the next digit:

poqpo

pqopo

Added up digit by digit, this will yield the result pqqoqo, which does indeed represent the sum of 75 and 57, as pqqoqo = 243 – 81 – 27 – 3 = 132.

In the triadic system, too, "one-multiplied-by-one" is even simpler than the "one-plus-one". It is based on the following nine very basic rules:

p · p = p	p · o = o	p · q = q
o · p = o	o · o = o	o · q = o
q · p = q	q · o = o	q · q = p

This is easy to commit to memory: multiplying by p leaves the cipher unchanged, multiplying by o yields o, and multiplying by q swaps the ciphers p and q and leaves o unaffected.

Moreover, we have to bear in mind that the multiplication by po, poo, pooo, poooo, … or, respectively, with qo, qoo, qooo, qoooo, … involves multiplication by p or q and the addition of one, two, three, four, … zeroes to the number. For example, to multiply the two numbers 75 = 1001011 and 57 = 111001, all that is required is to form the sum of the following numbers (because 57 = poooo + qooo + po):

(75 · poooo) = (poqpo · poooo) = poqpooooo

(75 · qooo) = (poqpo · pooo) = qopqoooo

(75 · po) = (poqpo · po) = poqpoo

This yields the result pqooqpoo, which indeed corresponds to 75 · 57 = 4275 because pqooqpoo = 6,561 – 2,187 – 81 – 27 + 9 = 4,275.

The most impressive advantage of the triadic system compared to binary is the extremely elegant way that it handles subtraction. Instead of having to learn a new set of calculation rules for "one-minus-one" all you need to remember is that subtraction is replaced by adding the negative value of the number to be subtracted. The negative value of any number is that number multiplied by q, which causes the ciphers p and q to replace one another (and leaves o unaffected). To subtract from 75 = poqpo the number 57 = pqopo, add –57 = (q · pqopo) = (qpoqo) to 75 = poqpo:

poqpo

qpoqo

The addition yields pqoo, which does indeed represent the difference between 75 and 57 as pqoo = 27 – 9 = 18. In addition to this, it is just as simple in the triadic system—as opposed to binary or base ten—to subtract the bigger number 75 = poqpo from the smaller number 57 = pqopo.

pqopo

qopqo

The result, qpoo, is the negative difference: qpoo = –27 + 9 = –18. As this simple example demonstrates, the triadic system does not need algebraic signs to differentiate between positive and negative numbers. Numbers whose highest place value is p are positive; numbers whose highest place value is q are negative. In other words: the triadic system has twice as many numbers (i.e., the positive and the negative ones) as binary and base ten, in which negative numbers require the algebraic sign – to be marked as such.

All this leaves a wide field for inconclusive speculation. Undeniably superior in many respects to the base ten system, triadic might have been adopted as standard if it had been discovered, say, by Leibniz, who would subsequently have been as enthusiastic about promoting its rise in the world as he was about promoting a reconciliation between the Protestant and the Catholic faiths. Hypothetical returns on investment in triadic might easily have proved more satisfactory than the ones he received in the pit of religion. We may go on to hypothesize that triadic would have spared generations of schoolchildren myriads of algebraic-sign errors, that bane of math classes, which continues to spread disaffection with the subject even to this day.

However, the die has been cast, and more than 300 years after Leibniz, it is hopeless to try and push triadic against a safely ensconced base ten.

74   ASCII is an acronym of American Standard Code for Information Interchange.

75   Examples of non-printable control characters in the ASCII code are the numbers 0000010 = 2 or 0000011 = 3, which mark the beginning and the end of a text, or the number 0001101 = 13, whose effect is comparable to the return key on a typewriter. Additional characters are such numbers as 0100101 = 37 for the percent sign (%), 0100110 for the ampersand (&) or 1000000 = 64 for the at sign (@).

76   The attached eighth bit makes a limited checking of the previous seven bits for transmission errors possible. While it cannot detect all possible mistakes or correct all the ones it does detect, it at least stops transmission and issues an error report. More sophisticated systems empower the check bit to carry out a more subtle analysis of transmission errors and sometimes also to correct them.

77   This is appended to the statement in a manner reminiscent of the ASCII code's check bit.

78   Subtracting the product of two bits from their sum corresponds in logical terms to linking two statements with the disjunctive *or* (not in the exclusive sense of either–or = Latin *aut–aut* but in the non-exclusive sense of Latin *vel*, as in the waiter's question "What are you going to eat or drink?"), known as a disjunction. Because of the equations

$$(1+1) - 1 \cdot 1 = 10 - 1 = 1 \qquad (0+1) - 0 \cdot 1 = 1 - 0 = 1$$
$$(1+0) - 1 \cdot 0 = 1 - 0 = 1 \qquad (0+0) - 0 \cdot 0 = 0 - 0 = 0,$$

the disjunction of two statements is true provided at least one of the statements is true; the disjunction of two false statements is false.

Subtracting a bit from 1 corresponds in logical terms to qualifying a statement through the word *not*, the negation. Because of $1 - 0 = 1$, the negation of a false statement is true, whereas $1 - 1 = 0$ demonstrates that the negation of a true statement is false.

Boole demonstrated in his mathematical logic how to break down all linguistic possibilities of linking up statements to conjunction, disjunction and negation. By doing so, he made these words amenable to binary processing.

79   This is why it is possible to copy CDs or digital video again and again with no perceptible loss in quality—something unheard of in the age of magnetic recording. After all, what is copied is the binary number on the disk. Even when copied for the thousandth time, a "0" will be a "0" and a "1" a "1".

80   To the left of the decimal point, in the units digit, is the integer quotient 1, followed after the decimal point by an 18-cipher block, 315 789 473 684 210 526, which is constantly repeated.

81   This term is used for situations in which computers are unable to process a command.

82   Eubulides, who lived in the fourth century BC, is credited with having thought up the "liar paradox", which Cicero hands down to us in his *Academica* as follows: "*Si te mentiri dicis idque verum dicis, mentiris an verum dicis?*" ("If you truthfully say you are telling a lie, are you in fact telling a lie or the truth?")

83   This is true of this type of curve. In the case of complex curves it is legitimate for a tangent to be in repeated contact with a curve without losing the title of tangent.

84   It is safe to assume that the coaches and spin doctors of all candidates work very hard to wean their charges from using numbers too frequently in discussions, particularly millions and billions, which are beyond the imaginative capacity of most of their viewers.

85   Relations between Leibniz and Newton were extraordinarily tense even though the two men never met. Newton suspected that Leibniz had stolen the seminal ideas from him that enabled him to develop calculus. Animosity ran high and the followers of the two great scholars organized themselves into hostile camps for a bitter war of words.

86   One of the problems that would make even the daemon halt in his tracks was identified by Laplace himself. If more than two atoms collide at the same time in the same spot, their subsequent trajectories cannot be determined. The supporters of the daemon tried to wriggle out of this by declaring the probability of this type of *triple hit* nil; we are assured it could never happen.

87   Fulfillment of this demand would embarrass the daemon itself. No one, not even the daemon, can meaningfully calculate with an *infinite number of decimal places*, if by this phrase we mean to include the whole infinite series of decimal places.

88   Many dice are wooden cubes with drill marks representing the "spots". The face with one spot is therefore slightly heavier than the opposite one with six spots; this affects the otherwise perfect symmetry of the die and marginally increases the probability of six turning up on top. "Sharpers" use loaded dice, in which weights or cavities wreak havoc with symmetry—and therefore probability.

89   That it is possible to formulate an answer in those terms has to do with the *law of large numbers*. Building on work done by the mathematicians Jakob Bernoulli (Swiss, 1654–1705) and Abraham de Moivre (French, 1667–1754), Laplace showed how, over a very large number of throws of the die, the incidence of a certain number of spots showing up on top was close to 16.67%; or

to be more precise, the incidence of a certain number of spots showing on top occurred closer to 16.67% the more often the die was thrown.

90 *Rouge* is the numbers 1, 3, 5, 7, 9, 12, 14, 16, 18, 19, 21, 23, 25, 27, 30, 32, 34 and 36. In addition to *rouge* and *noir*, distinctions are made between *manque* (the numbers 1 to 18) and *passe* (the numbers 19 to 36) and between *pair* (even numbers) and *impair* (odd numbers). These are only a few of the many gaming possibilities offered by roulette. The choice of multiple-divisible 36 for the number of pockets perhaps reflects the friendly terms on which one of the game's putative inventors, Blaise Pascal, was with numbers.

91 This is Laplace's general definition of probability: it is the ratio between *favorable outcomes* and the sum total of *possible outcomes* (an outcome being the result of an observed random process). Individual outcomes are perfectly symmetrical with regard to the process so that the same probability can be guaranteed for the different outcomes—as is the case with the six faces of a die, the two sides of a coin, the concealed playing cards of a set of cards, the pockets at the bottom of the roulette cauldron etc.

92 It is indeed probable that the sun will not rise again one morning at some stage inside the next 10 billion years because it will by then have poured all of its energy out into space.

93 Because this trial failed to recruit same-sized groups of respondents for each stage, these numbers, obtained at great pains among the community of migraine sufferers, have no significance whatever. Among statisticians, the effect is known as *Simpson's Paradox*.

94 To mention only three: the 1954 classic, *How to Lie with Statistics*, by Darrell Huff, after which the whole genre has now been named; the Mark Twain-inspired title *Damned Lies and Statistics: Untangling Numbers from the Media, Politicians, and Activists* by Joel Best and *A Mathematician Reads the Newspaper* by John Allen Paulos.

95 The reason for this was already evident to Laplace. In every game, there are 37 pockets in which the ball can land. In ten games, the number of "possible outcomes" equals

$$37^{10} = 37 \cdot 37 \cdot 37 \cdot 37 \cdot 37 \cdot 37 \cdot 37 \cdot 37 \cdot 37 \cdot 37 = 4,808,584,372,417,849.$$

This is the sum total of possibilities to compile lists of numbers between 0 and 36 each containing 10 entries. From the sum total of *possible* cases, only those are regarded as *favorable* in which no number is listed more than once. For the first number on that list there are 37 permissible entries; for the second one, there are only 36—the first number to have been entered now being excluded; for the third one, there are only 35—the first two numbers to have been entered now being excluded, and so on. The number of possibilities to compile lists of numbers ranging from 0 to 36 in such a way that no number occurs more than once is

$$37 \cdot 36 \cdot 35 \cdot 34 \cdot 33 \cdot 32 \cdot 31 \cdot 30 \cdot 29 \cdot 28 = 1,264,020,397,516,800,$$

this is the number of favorable cases. The ratio between favorable cases and the sum total of possibilities,

$$1,264,020,397,516,800/4,808,584,372,417,849 \sim 0.263 = 26.3\%,$$

indicates how surprisingly low the probability is for the ball to land in ten different pockets in ten consecutive games.

96   It is even more surprising how high the probability is of finding two persons in a random group of 36 who share the same birthday (we will assume for the sake of simplicity that those born on February 29 celebrate their birthday in non-leap years on February 28. Even though the number of people we are considering is less than a tenth of the number of days per year, the probability is an amazing 83.22%. The numbers used in this calculation, 365^{36} (multiplying 365 36 times with itself), result in a truly mind-boggling 93 digits.

97   The same lack of mathematical imagination accounts for the reactions to an experiment that the Austrian mathematician Walter Schachermayer (b. 1950) loved to conduct with his students. He asked a group to flip a coin 200 times and to jot down the results of head or tail using 0 and 1. A control group was asked to compile at random a list of the numbers 0 and 1 consisting of 200 entries. When the two lists were presented to him he did not normally need to be told which group had compiled which list: he knew that the list consisting of the numbers made up at random always contained less long repetitive series of 0s and 1s than the list of the coin-flippers. For roulette aficionados, this does not come as a surprise. They know from experience that there will be long series of the ball ending up on *rouge* or *noir*, respectively. To refer to this fact as evidence of a *law of averages* is nonsensical; serial occurrences are the result of chance, which cannot be distilled into forecasts.

      Further evidence of how alien very large numbers are to our way of thinking can be obtained from an analysis of the principles governing lotteries. In Austria's "6 out of 45", where a ticket consists of six numbers between 1 and 45, there are 8,145,060 possibilities per ticket. (There are $45 \cdot 44 \cdot 43 \cdot 42 \cdot 41 \cdot 40$ = 5,864,443,200 possible lists consisting of six different numbers between 1 and 45. Because the point of the lottery is guessing the correct numbers regardless of the order in which they come up and because there are $6 \cdot 5 \cdot 4 \cdot 3 \cdot 2 \cdot 1 = 720$ different ways of ordering these six numbers, the quotient of 5,864,443,200/720 = 8,145,060 indicates the number of possible different tickets.) This shows considerable foresight on the part of the organizers of the lottery, because Austria has approximately 8 million inhabitants. The lottery caught on beyond expectations, and more than 8 million tickets per round are usually sold. Once, when a triple jackpot was played out, the number rose to an extraordinary 52.5 million. In spite of this, jackpots and rollover jackpots are common. How do we account for this phenomenon?

      The solution of the riddle may well be that an unexpectedly large number of players bet on the same six numbers without being aware of how many others are playing the same numbers. It is quite unbelievable how many players bet on the sequence 1, 2, 3, 4, 5, 6. As a bet, this sequence is equally probable to come up as any other selection of six numbers between 1 and 45. When it does come up one day, the winners will be amazed how many will share the first prize. It certainly appears that lotto players find it difficult to visualize the enormous variety offered by 8,145,060 possibilities. They stick to the same patterns of selecting six numbers from 1 to 45 and prefer "lucky" numbers partly for superstitious, partly for wholly irrational reasons.

If you really want to play the lottery, you should at least try to go for numbers that other players tend to leave aside. If you do win a first prize that way—the probability for that outcome, calculated by $1/8{,}145{,}060 \approx 0.0000123\%$, being only fractionally above zero—you can count on winning a really large sum of money. A method that would preclude being influenced by lucky numbers or geometrical patterns might, for example, involve writing the numbers from 1 to 45 on playing cards, shuffling them carefully and betting on the numbers you draw from the pack. The organizers of the lottery offer a service, in which a random number generator compiles a ticket for the player, but this anonymous procedure probably destroys for many the illusion that the lottery can be relied on to deliver with a high degree of probability; namely, that of having unlimited wealth at the tip of your pen...

98   This is the case because the area of a bar below the income distribution graph represents the total number of employees who draw a net income within the specified bracket.

99   Admittedly, sulphur immediately combines with oxygen from the air to form sulphur dioxide.

100 The nucleus later also turned out to consist of particles: protons (whose numbers are identical with the atomic number) and neutrons (which approximate protons in weight and contribute to the density of the nucleus).

101 Perhaps it could also be quarks, which are thought to be the constituent elements of protons and neutrons today.

102 This photoelectric effect has been harnessed for a broad range of mundane technological applications today, from recording equipment to door openers etc.

103 Even in a truncated form this fraction still yielded the complex ratio of 68,547:50,777.

104 The Pythagoreans called this the *technique of alternate elimination.*

105 It was one of Bohr's strokes of genius to relate the value of the Rydberg constant to a number of measurements carried out by his predecessors, including that of the charge of the electron and its mass, the speed of light and Planck's constant.

106 A long-lived fairy story centering on "Bohr's model of the atom" has him seeing electrons as tiny spheres circling around the nucleus in allowed orbits. This flight into model-based interpretation of the atom as a miracle of precision engineering is completely alien to Bohr's intentions.

In his wonderful *Geschichte der Quantentheorie*, Friedrich Hund describes the genesis of Bohr's epoch-making treatise of the summer of 1913:

> In the summer of 1912, Bohr had joined Rutherford, after a short and somewhat disappointing stay with J. J. Thomson at Cambridge, for a round of experiments in Manchester. What he found fascinating about Rutherford's model was the idea that all physical and chemical properties of a chemical element could possibly be the consequence of one single number, the number of elementary charges in the nucleus (which is identical with the number of electrons). Yet he instantly realized that Rutherford's model could not account for the atoms' stability. There

are fragments of a first draft from the summer of 1912 in which Bohr underlines that no length had been distinguished, that the movements of the electrons were not stable and that the stability issue had to be tackled in a totally different manner. ... A mechanical explanation was a non-starter. The chemical properties and the periodic table of elements were to be explained through rings of electrons with the decisive numbers of electrons being positioned in the outermost circle, which was of course viewed differently by J. J. Thomson. Bohr also promises an explanation of the periodically oscillating atomic volume of the elements, of the dependence of some properties on the atomic number and an explanation of the stability of chemical compounds. Spectroscopy is not mentioned.

It would appear from Bohr's letters that as late as February 1913 he had still not included the spectral laws in his deliberations. Yet at the beginning of March he sent the first parts of the treatise to Rutherford; the letter accompanying it claims that an explanation has been found for the hydrogen spectrum and for the significance of the Rydberg constant. ... In other words, Bohr developed the theory of the hydrogen spectrum in less than a month. He later said: The moment I saw Balmer's formula, everything became clear instantly.

107 This is because, in addition to its energy content, the entire *rotary impulse* of the electron—that is, the impulse that makes it revolve in its orbit added to its spin—turned out to be significant features of the electron in the atom in terms of quantum theory and must therefore be expressed through quantum numbers. For undergraduates in a quantum-theory course, it is a matter of consternation that reference to the *rotary impulse* of the electron in the atom does not justify the idea of the electron orbiting around the nucleus like a satellite. In the same way, references to the electron's *spin* have nothing whatever to do with the graphic idea of the electron revolving about its own axis.

108 We owe this seminal insight to Immanuel Kant: "Our senses do not deceive us. This sentence is a rejection of the most important, and at the same time most trivial reproach that can be leveled at the senses: not because they always tell us the truth, but because they do not tell us anything at all: it is the intellect that misunderstands, not the senses." (Immanuel Kant, "Anthropologie in pragmatischer Hinsicht." In *Werkausgabe*, edited by Wilhelm Weischedel, vol. XII, p. 435f. Frankfurt/Main: Suhrkamp, 1980.)

109 *Tohu va-bohu* is the Bible's word for "chaos"; in their exemplary German translation of the Hebrew Bible, Buber and Rosenzweig use the phrase "*Irrsal und Wirrsal*", error and confusion.

110 Whether the modern cosmological legend is preferable to other creation myths is a moot point.

111 When the play was first performed in Paris in 1945, critics were quick, in the light of contemporary political developments, to draw parallels between the two potential world-wreckers Caligula and Hitler. However, Camus equipped Caligula with intellectual powers far superior to his counterpart so that the parallel is only a superficial one. This is a very disquieting thought as regards to the future. As Carl Amery has pointed out in his book, *Hitler as Forerunner*, it is by no means inconceivable for a Third-Reich-style dictatorship skillfully concealing the formerly evident crass lack of professionalism with a varnish of scientific glamour and scientific vocabulary to be revived in the West in the

foreseeable future. As Caligula is struck down by the conspirator's daggers, he calls out to them and to the audience: "I am not dead yet."

112 Another, perhaps less dramatic example is Kant's claim that Euclid's geometrical axioms were the only basis that would make an understanding of space possible. The discovery of non-Euclidean geometries (in one of whose versions, there are no such things as parallels; whereas in the other, infinitely many parallels can be drawn to a given straight line through a point outside that line) seems to disprove Kant. The contradiction gained in significance when Einstein proposed, as an aspect of the general theory of relativity, that non-Euclidean geometries featured in the structure of the universe.

113 Karl Jaspers was the first to point out that the misconceived term *"Verbrechen gegen die Menschlichkeit"* ("crimes against humane-ness") should be replaced by the much more apposite *"Verbrechen gegen die Menschheit"* ("crimes against humanity").

114 In an article published by Süddeutsche Zeitung on August 18, 2003, Dorothea Razumovsky recalls a conversation between Theodor Adorno, Max Horkheimer and Andreas Razumovsky, her future husband, which took place in her presence one evening toward the end of November or December 1961, when Eichmann's abduction from Argentina by a Mossad special unit elicited general admiration:

> Of course all of us had read Hannah Arendt's reports on the trial in the "New Yorker"—the book on "The Banality of Evil" did not get published until two years later—and we had been following developments in Jerusalem on a day-to-day basis. Now however sentencing through the three judges was imminent. Was he going to be sentenced to death or not?
>
> Agreement was soon reached among us that it would be a tremendous burden on the young state of Israel to resort in its turn to violence. Yet the reasons given for that opinion could hardly have been more divergent. Adorno mentioned the duty of the court to establish the individual guilt of the accused. An integral part of this was the question as to his personal culpability: was it possible for someone who had been part of a psychopathic system which was not only tolerant of mass killings but purposefully organized them itself, a system moreover that had discarded all moral values, the Ten Commandments included, and had been backed in this by the majority of the population, was it really possible for such a person objectively to differentiate between good and evil? Punishment as deterrent was out of the question in this case and punishment as expiation problematic. What had been so dehumanized about the Nazi regime was its bureaucratic administration of genocide, without any human emotions, without any personal hatred.
>
> Horkheimer, who happened to be present, took a different line altogether. He sat there for a long time without a word, his leonine head cast back in a pose of thoughtfulness, the mouth open, almost as if he was in a reverie, until he contradicted his friend vehemently. This was not an individual that was being tried, he said. No, not even the Nazi regime but anti-Semitism as such, and not only the modern variant but anti-Semitism through the ages. It was not sufficient to punish this individual criminal because his execution might suggest that justice had been done and the case was closed. One had to regard Eichmann as a symbolic figure, whose forehead should be branded with the sign of Cain. And then let him be left to his own devices like a mangy dog.

This idea was so convincing and so powerful that we found ourselves agreeing with it. Subsequently I became less sure and regretted above all that I had not written it all down after I got home. For, as I understood only as time passed, something essential had revealed itself in the discussion: Horkheimer articulated in his words the ethics of the Old Testament, Adorno that of the New Testament or, rather, of the Christian-Occidental tradition.

115 It is not only aesthetic standards that are beyond objective analysis, however absolute they may appear to the individual; moral standards are equally elusive. Kant's attempt to formulate a universally binding categorical imperative never got beyond the realm of the formal. Virtues such as responsibility, love or hope, even though they may seem absolutely cogent categories to us, refuse to be harnessed in a straightforward moral code.

116 The process of counting as a project that is incommensurate with our finite resources is an intriguing idea of the contemporary Dutch mathematician Wim Veldman (b. 1946).

117 The fact that all mathematical deliberations require the two ideas of defining a starting point and of then proceeding step by step was first recognized to be generally valid by Pascal, who formulated it as the *principle of complete induction*. It is nevertheless worth pointing out that *complete induction* had implicitly been used already by the Pythagoreans and, as the example outlined here proves, by other mathematicians in antiquity such as Archimedes.

118 The square of the side length of the polygon with twice the number of vertices is calculated from the side length of the originally given polygon inscribed in a circle with radius = 1 unit as follows: subtract the square of the side length of the originally given polygon from 4 and take the square root of the difference; then subtract this root from 2. (This example illustrates how much more elegantly a complex state of affairs can be expressed through a formula. However, neither Archimedes nor Pascal used formulae of the type that is common today but clothed their insights in words rather like the ones above.)

119 A second example concerns the ratio between the diagonal of a square and its sides, which is represented by $\sqrt{2}$, the square root of 2. When multiplied with itself, this will yield the result: 2. This ratio, calculated to the first ten decimal digits, $\sqrt{2} = 1.4142135623\ldots$

A third example concerns $\sqrt[12]{2}$, the twelfth root of two, which signifies the interval of the minor second in Simon Stevin's equal temperament, a value which yields 2 when multiplied twelve times with itself. Calculated to the first ten decimal digits, $\sqrt[12]{2} = 1.0594630943\ldots$

A fourth example is the ratio of the longest side of that right triangle to its shortest side in which the ratio between the second longest side to the shortest is exactly 3. As can be demonstrated by means of Bhaskara's figure, that longest side squared includes ten squares with the shortest side as side length. This ratio is therefore called $\sqrt{10}$, the square root of 10. Multiplied with itself it will yield the result: 10. Calculated to the first ten decimal digits, $\sqrt{10} = 3.1622776601\ldots$

A fifth example, the oldest in historical terms, dates back to the Pythagoreans. One of them, Hippasos of Metapont, realized that the diagonals of a

regular pentagon, which form the so-called pentagram, intersect in a remarkable way. Four sides can be lifted from the diagonal with its two endpoints
and the two points of intersection. The ratio between the lengths of the longest of these four sides—the diagonal—and the second longest side—the side
from the endpoint to the farther point of intersection—will be identical with
the ratio between the second longest to the third longest side, the one from
the endpoint to the nearer point of intersection. The same applies to the ratio
between the third longest to the shortest side, i.e. the one that connects the two
points of intersection. This truly miraculous ratio has been called the *Golden
Mean* since the time of Kepler; its symbol is the Greek letter ϕ. Calculated to
the first ten decimal digits, the ratio of the *sectio aurea* = 1.6180339887….

In all of these examples, the infinite is concealed after the three dots at the
end of the ten decimal places; without surcease, cipher follows upon cipher in
an incomprehensible, jumbled sequence.

120 The so-called periodic decimal numbers, which result as quotients from such
divisions as 1/3 = 0.333333333… or 22/7 = 3.142857142857142857…, may be
compared to crystals in nature. In crystals, too, there is periodicity in the repetition of the same molecular structures. Seen in that light, values such as π or
"aleatoric values" correspond in nature to "aperiodic crystals", a term which
was proposed by Erwin Schrödinger in his *What is Life*; it reflects his anticipation of the structure of nucleic acid in the living cell.

121 "Algorithm", a portmanteau from the Greek *árithmos*, meaning "number",
and the name of the Persian mathematician Al Khwarizmi, signifies a complex
operation that unfolds according to clearly stated, predefined rules.

122 It is by no means far-fetched to draw a parallel between Dedekind's hyper-
bolic claims and the hubris and complete breakdown of self-control of Camus'
Caligula.

123 The first to formulate this idea was the Greek thinker Anaximander (c. 610–
c. 546 BC). For him, the infinite, which he called *to ápeiron*, is an endless mass
without limits, in which everything that exists has its origin.

Acknowledgments

Pictures in a book like this are oases of color and concreteness, punctuating an arduous journey through a desert of print. I am deeply grateful to Erich Lessing for generously supplying me with a multitude of beautiful photographs from his apparently inexhaustible treasure house. His photographs are tangible proof of how numbers may serve as a link between pedestrian mathematics and the most sublime works of art.

I also want to express my gratitude to the team at the Picture Archive of the Austrian National Library under its director, Hans Petschar, for supporting me in my research with their expert knowledge. A considerable number of colleagues have also contributed in various ways to the genesis of this book; I have thanked many of them by name in the German edition, and I would like to repeat those thanks at least in a summary fashion here.

The eight chapters of the book are based on material that first saw the light of day as presentations and workshops on these eight topics which I delivered in Vienna at *math.space*. An integral part of MuseumsQuartier, itself one of the ten largest museum complexes worldwide, *math.space* is financed by the Ministry of Education of Austria with support from private sector sponsors. It was founded with the intention of increasing public awareness of the eminent cultural achievement that is mathematics. The fact that attendance has doubled every year since *math.space* was founded in 2001 is a matter of great pride to all who are involved in this thriving enterprise, and I consider it a privilege to have contributed through the present book.

A book's success is, of course, owed at least partly to the publishers. "*Merci cordialement,*" therefore, to Ulrike Schmickler-Hirzebruch of the publishing house of Vieweg, who has expertly guided the book through four reprints, and to Alice and Klaus Peters, who have decided to offer the book a home in the United States.

David Sinclair-Jones and Otmar Binder are a team of translators of which authors can only dream. In being transformed into *Numbers at Work*, *Der Zahlen gigantische Schatten* has benefited substantially from the countless suggestions that they made. My debt of gratitude to them is very great indeed.

Lastly, I want to say that without the help, encouragement and loving forbearance I received throughout from my wife Bianca and our children, this book could never have been written.

Figure Credits

1 Pythagoras of Samos, 569(?)–475(?) BC; mezzotint engraving by Johann Faber of Peter Paul Rubens's drawing of an antique marble bust. Picture Archive of the Austrian National Library.

2 Thales of Miletus, 624(?)–547(?) BC; anonymous engraving. Picture Archive of the Austrian National Library.

3 Photograph of a solar eclipse; Munich Astronomical Observatory 1905. Erich Lessing Culture and Fine Arts Archives 24–01–01/50.

4 Stone tablet showing the calculation of the area of a plot of land near Umma, Mesopotamia (Iraq); Louvre, Departement des Antiquités Orientales. Erich Lessing Culture and Fine Arts Archives 08–02–06/12.

5 Geometry room in the Collegium Maius (of the old university); Collegium Maius, Cracow, Poland. Erich Lessing Culture and Fine Arts Archives 24–01–01/4.

7 Michelangelo Buonarotti, 1475–1564: The Holy Family; tempera, 1504–1506; inventory # 1456, Uffizi, Florence. Erich Lessing Culture and Fine Arts Archives 40–08–18/63.

8 Lodovico Buti, 1560–1603: Abraham is visited by three angels; inventory # 1520, Museum of Art History Vienna, Gallery of Paintings. Erich Lessing Culture and Fine Arts Archives 40–08–07/58.

9 Bruno Girin, DHD Multimedia Gallery.

10 Heliocentric armillary sphere with the twelve signs of the zodiac on the outermost ring; Italian workmanship, unsigned, around 1810; private collection, Vienna. Erich Lessing Culture and Fine Arts Archives 32–01–01/52.

12 Aachen cathedral. The octagonal basilica was built by Charlemagne shortly before 800. The chandelier inside the octagon is a gift from the Emperor Friedrich Barbarossa, 1125–1190. Erich Lessing Culture and Fine Arts Archives 16–01–03/1.

15 Notre Dame de Paris. The first church was built on the foundations of a Roman temple dedicated to Jupiter; the present-day cathedral was started in 1163 and finished in 1330. Erich Lessing Culture and Fine Arts Archives 15–03–09/67.

17 Albrecht Dürer, 1471–1528: Melencolia I; copper engraving, 1514. Albertina, Vienna.

19   Painter of the French School: Jacob's Ladder; Musée du Petit Palais, Avignon, France. Erich Lessing Culture and Fine Arts Archives 40–12–20/6.

21   Hans Acker, 16th century: Moses on the mountain; window of the Besserer Chapel of Ulm cathedral, Ulm. Erich Lessing Culture and Fine Arts Archives 16–01–02/46.

24   Elias Gottlob Haussmann, 1695–1774: Johann Sebastian Bach, 1685–1750; oil painting, 1764. Museum of the Fine Arts, Leipzig. Erich Lessing Culture and Fine Arts Archives 40–06–02/4.

26   Jean Baptiste Joseph Baron de Fourier, 1768–1830; lithograph by Jules Boilly, 1823. Picture Archive of the Austrian National Library.

34   Leonhard Euler, 1707–1783; mezzotint engraving by Johann Stenglin, 1768. Picture Archive of the Austrian National Library.

37   Simon Stevin, 1548/49–1620; portrait in the library of Leiden University, Holland. Austrian Central Physics Library.

40   Constructed by Harald Gert Tranacher, Vienna, on the basis of material supplied by the author.

41   Hugo von Hofmannsthal, 1874–1929; Photographie Wasow, Munich. Picture Archive of the Austrian National Library.

42   Parmenides of Elea, born c. 515 BC; engraving by Guglielmo Morghen. Picture Archive of the Austrian National Library.

43   A trompe l'oeil gate near the parish church of Ingolstadt, Bavaria. Erich Lessing Culture and Fine Arts Archives 15–04–07/56.

44   Aristotle, 384–322 BC; engraving by Samuele Jesi. Picture Archive of the Austrian National Library.

45   Sunrise. View from Mount Sinai, where Moses received the Ten Commandments. Erich Lessing Culture and Fine Arts Archives 08–03–01/25.

46   Part of a planetarium showing Sun, Moon and Earth; Italian workmanship, 18th century; Università di Bologna, Italy. Erich Lessing Culture and Fine Arts Archives 32–01–01/59.

48   Gaius Julius Caesar, 100(?)–44 BC; engraving by Pierre Daret after an antique coin. Picture Archive of the Austrian National Library.

49   Johannes Regiomontanus, 1436–1476; engraving by Georg Wolfgang Knorr. Picture Archive of the Austrian National Library.

50   Pope Gregory XIII, 1502–1585. Picture Archive of the Austrian National Library.

51   Armillary sphere with clock, 1572; Rosenborg Castle, Copenhagen, Denmark. Erich Lessing Culture and Fine Arts Archives 30–01–04/47.

52   Christiaan Huygens, 1629–1695; engraving by G. Edelinck. Picture Archive of the Austrian National Library.

54   Albert Einstein, 1879–1955; photograph by T. Fleischmann, New York, 1946. Picture Archive of the Austrian National Library.

55   A moving clock runs slower than a stationary one. Erich Lessing Culture and Fine Arts Archives 24–01–01/56.

56   Ludwig Boltzmann, 1844–1906; lithograph by Rudolf Fenzl, 1898. Picture Archive of the Austrian National Library.

57   Marc Chagall, 1889–1986; Mountebank, 1943. The Chicago Art Institute, Chicago. Erich Lessing Culture and Fine Arts Archives 40–12–17/23.

58   Damon Hart-Davis, DHD Multimedia Gallery.

59   Erwin Schrödinger, 1887–1961; Photographie Landesbildstelle Berlin. Picture Archive of the Austrian National Library.

60   Nicolas of Verdun, c. 1150–1205: The Birth of Christ, from the Verdun Altar (after 1181); collections of the monastery of Klosterneuburg, Austria. Erich Lessing Culture and Fine Arts Archives 15–01–01/43.

61   God surveys the universe; illuminated Bible, probably from Reims, middle of the 13th century; Cod. 2554 fol. I; Austrian National Library. Erich Lessing Culture and Fine Arts Archives 15–02–04/41.

62   Frans Hals, 1581–1666; René Descartes, 1596–1650; oil painting after a lost original by Frans Hals. Departement des Peintures, Louvre, Paris. Erich Lessing Culture and Fine Arts Archives 26–03–02/38.

63   Giordano Bruno, 1548–1600; engraving by Guglielmo Morghen after a drawing by Aniello d'Aloisio. Picture Archive of the Austrian National Library.

64   Egyptian scribes; detail from a wall painting in the tomb of Mennah, surveyor and civil servant under Pharaoh Thutmosis IV. (18th dynasty, 16th–14th century BC), in the necropolis of Sheikh Abd al-Qurnah, Luxor-Thebes. Erich Lessing Culture and Fine Arts Archives 08–01–01/49.

69   The terrace of Belvedere Castle (Prague, Czech Republic), where Tycho de Brahe and Johannes Kepler carried out astronomical observations; the quadrant comes from Habermehl, Dresden. Erich Lessing Culture and Fine Arts Archives 24–01–02/22.

71   Justus (Joos) van Gent, 1435(?)–1480 (?): Claudius Ptolemy, 2nd century AD; from a series of 28 portraits painted around 1475 for Federico da Montefeltro of Urbino. Erich Lessing Culture and Fine Arts Archives 40–03–05/16.

73   Gnomonic representation of the hemisphere with Syene, Egypt, as the point of map origin. Hans Havlicek, Vienna University of Technology.

74   NASA National Aeronautics and Space Administration.

75   Pierre de Fermat, 1601–1665; anonymous engraving. Picture Archive of the Austrian National Library.

76   Aristarchus of Samos, around 320 BC, and Hipparchus of Nikaia, second century BC, in an allegorical dispute about the geocentric and the heliocentric systems; engraving from the Lavater Collection. Picture Archive of the Austrian National Library.

77   Johannes Kepler, 1571–1630; anonymous oil painting, 1627; Musée de l'Œuvre Notre-Dame, Strasbourg, France. Erich Lessing Culture and Fine Arts Archives 24–01–02/18.

79   Painter of the German School, 1575: Nicolaus Copernicus, 1473–1543. Legend: "Clarissimus et Doctissimus Doctor Nicolai Copernicus... Astronomus Incomparabilis" (The renowned and most learned Doctor Nicolai Copernicus...

Astronomer extraordinary); Collegium Maius, Cracow, Poland. Erich Lessing Culture and Fine Arts Archives 24–01–01/1.

80   Nicolaus Copernicus' treatise "De Revolutionibus Orbium Coelestium libri sex", 1543; Library of the Collegium Maius, Cracow, Poland. Erich Lessing Culture and Fine Arts Archives 24–01–01/26.

81   Friedrich Wilhelm Bessel, 1784–1846; photoengraving of a painting dated 1839. Picture Archive of the Austrian National Library.

83   NASA National Aeronautics and Space Administration.

84   Rainer Maria Rilke, 1875–1926; photograph, 1906. Picture Archive of the Austrian National Library.

86   NASA National Aeronautics and Space Administration.

88   NASA National Aeronautics and Space Administration.

93   Gottfried Wilhelm Leibniz, 1646–1716; engraving by Johann Benjamin Brühl. Picture Archive of the Austrian National Library.

94   Thomas Hobbes, 1588–1679; stipple engraving by James Posselwhite, c1840, after a painting by William Dobson. Picture Archive of the Austrian National Library.

95   Reliquiary in the form of a bust of Charlemagne. The reliquiary contains fragments of Charlemagne's skull; it was commissioned by the Holy Roman Emperor Charles IV in 1349 and featured in his coronation in 1355; Aachen Cathedral Treasury. Erich Lessing Culture and Fine Arts Archives 15–01–04/2.

96   La prise de la Bastille, le 14 juillet 1789 – The Storming of the Bastille on July 14, 1789; private collection, Vienna. Erich Lessing Culture and Fine Arts Archives 39–15–02/4.

97   Bavarian and French troops near Vimy, 1914; Bavarian Army Museum, Ingolstadt, Bavaria. Erich Lessing Culture and Fine Arts Archives 17–02–01/15.

98   Budget Stock Photos.

100 George Boole, 1815–1864; xylograph, 1865. Picture Archive of the Austrian National Library.

101 William Shakespeare: Romeo and Juliet, with Johannes Krisch as Romeo, Eva Herzig as Juliet and Bernd Birkhahn as Friar Laurence; from a production at the Burgtheater, Vienna; photograph by Andreas Pohlmann, Munich.

102 William Shakespeare: Romeo and Juliet, with Michael Rotschopf as Tybalt, Walter Wilke (Benvolio), Markus Hering (Mercutio), Johannes Krisch (Romeo), Harald Höbinger (Balthasar) and Haymon Maria Buttinger as one of the musicians; from a production at the Burgtheater, Vienna; photograph by Andreas Pohlmann, Munich.

103 Kurt Gödel, 1906–1978; photograph. Austrian Central Physics Library.

104 Luitzen Egbertus Jan Brouwer, 1881–1966; photograph by E. van Moerkorken, 1943.

108 Pierre Simon, Marquis de Laplace, 1749–1827; ad vivum drawing by Jean Baptiste Regnault. Picture Archive of the Austrian National Library.

109 Sir Godfrey Kneller, 1646–1723: Sir Isaac Newton, 1642–1727; Trinity College, Cambridge, United Kingdom. Erich Lessing Culture and Fine Arts Archives 24–01–01/27.

110 NASA National Aeronautics and Space Administration.

113 Werner Heisenberg, 1901–1976; intaglio print from "Deutsche Illustrierte" no.14. Picture Archive of the Austrian National Library.

114 Detail from Oswald Skene, DHD Multimedia Gallery.

115 NOAA National Oceanic and Atmospheric Administration.

116 Adrienne Hart-Davis, DHD Multimedia Gallery.

117 Edvard Munch, 1863–1944: Roulette Table; oil painting, 1891–1892. Munch Museum, Oslo. Erich Lessing Culture and Fine Arts Archives 40–17–03/34.

118 Sunrise on the western shore of the Sea of Galilee (Yam Kinneret), Bethsaida, Israel. Erich Lessing Culture and Fine Arts Archives 08–03–04/29.

119 NASA National Aeronautics and Space Administration.

126 "Voltaire" (François-Marie Arouet), 1694–1778; engraving by Eugène Magne. Picture Archive of the Austrian National Library.

127 Niels Bohr, 1885–1962; photograph by Jacobsen, Copenhagen, 1951. Picture Archive of the Austrian National Library.

128 Morguefile mconnors.

129 Budget Stock Photos.

130 Budget Stock Photos.

132 Fully functional toy steam engine, c. 1860; Museum Carolino Augusteum, Salzburg, Austria. Erich Lessing Culture and Fine Arts Archives 30–01–06/64.

133 Robert Boyle, d. 1621; mezzotint engraving by Pieter Schenk after a painting by J. Kerseboom. Picture Archive of the Austrian National Library.

134 Sir Joseph John Thomson, 1856–1940; copper intaglio print after a photograph. Picture Archive of the Austrian National Library.

135 Cathode rays deflected by electromagnetic fields. ETH Zurich, Switzerland. Erich Lessing Culture and Fine Arts Archives 24–01–01/53.

137 Sir Ernest Rutherford, 1871–1940; copper intaglio print after a photograph. Picture Archive of the Austrian National Library.

138 Perspex model of uranium-235 atom commissioned by the Union Carbide Corporation, New York. Erich Lessing Culture and Fine Arts Archives 24–01–01/52.

141 Morguefile mconnors.

144 Gustav Robert Kirchhoff, 1824–1887, and Robert Wilhelm Bunsen, 1811–1899; Photographie Lange, Heidelberg. Picture Archive of the Austrian National Library.

148 Victor Frederick Weisskopf, 1908–2002; photograph, Visual Education Service, University of British Columbia, Vancouver, B.C., 1957. Picture Archive of the Austrian National Library.

149 Carl Friedrich Freiherr von Weizsäcker, b. 1912; Photographie Kempe; Staatliche Landesbildstelle Hamburg, 1964. Picture Archive of the Austrian National Library.

150 François Quesnel II, 1637–1699: Blaise Pascal, 1623–1662; Musée National du Château, Versailles, France. Erich Lessing Culture and Fine Arts Archives 26–03–09/17.

151 Sandro Botticelli, 1445–1510: detail from The Birth of Venus; tempera c. 1486; inventory # 878, Uffizi, Florence. Erich Lessing Culture and Fine Arts Archives 40–07–10/11.

152 Philip J. Davis, b. 1923; photograph by Christa Binder, Vienna.

153 NASA National Aeronautics and Space Administration.

154 Albert Camus, 1913–1960; photograph by Bernhard, Paris, 1953. Picture Archive of the Austrian National Library.

156 Emily Dickinson, 1830–1886. University of Illinois at Urbana-Champaign (available from Wikimedia Commons).

157 William Turner, 1775–1851: Rain, steam and speed; oil painting, before 1844, NG 538, National Gallery, London. Erich Lessing Culture and Fine Arts Archives 40–06–01/15.

158 Roman Opalka, b. 1931: Opalka 1965 1–∞; detail 4 185 294–4 207 974. MUMOK, Museum Moderner Kunst Stiftung Ludwig, Vienna.

161 Archimedes, 287 (?) –212 BC; after an oil painting by Domenico Fetis. Picture Archive of the Austrian National Library.

162 Hermann Weyl, 1885–1955; photograph. Austrian Central Physics Library.

163 Robert (Edler von) Musil, 1880–1942. Picture Archive of the Austrian National Library.

All other pictures supplied by the author.

Bibliography

PYTHAGORAS: NUMBERS AND SYMBOL

A. Aaboe. *Episodes from the Early History of Mathematics.* Random House, 1964.

W. S. Andrews. *Magic Squares and Cubes.* Cosimo, 2004.

C. Balmond. *Number 9.* Prestel, 1998.

A. H. Beiler. *Recreations in the Theory of Numbers.* Dover, 1964.

E. T. Bell. *The Magic of Numbers.* Dover, New York, 1991.

E. Bischoff. *Mystik und Magie der Zahlen.* Fourier, 3rd ed. 1997

P. Daniel. *En-Sof.* Edition Splitter, 1992.

J. J. Davis. *Biblical Numerology: A Basic Study of the Use of Numbers in the Bible.* Baker Academic, 1968.

M. Gell-Mann. *The Eightfold Way: A Theory of Strong Interaction Symmetry.* DOE Technical Report, March 15, 1961.

J. Godwin. *The Harmony of the Spheres: The Pythagorean Tradition in Music.* Inner Traditions, 1992.

W. K. C. Guthrie. *The Earlier Presocratics and the Pythagoreans.* Cambridge University Press, 1979.

A. Peter Hayman (translator and editor). *Sefer Yesira.* Texts and Studies in Ancient Judaism 104. Mohr Siebeck, 2004.

V. F. Hopper. *Medieval Number Symbolism: Its Sources, Meaning, and Influence on Thought and Expression.* Dover, 2000.

G. Ifrah. *The Universal History of Numbers: From Prehistory to the Invention of the Computer.* Wiley, 1999.

A. Kircher. Arithmologia sive de abditis numerorum mysteriis. Rome, 1665.

G. Locks. *Spice of Torah-Gematria.* Judaica Press, 1985.

M. Maimonides. *Mishneh Torah.* Edited and translated by M. H. Hyamson. New York, 1937.

K. Menninger. *Number Words and Number Symbols: A Cultural History of Numbers.* Dover, 1992.

A. Schimmel. *The Mystery of Numbers.* Oxford University Press, 1993.

T. Schirrmacher. *Verborgene Zahlenwerte in der Bibel.* VKW, 2001.

D. Wells. *The Penguin Book of Curious and Interesting Numbers.* Penguin Books, 1986.

H. Weyl. *Symmetry.* Princeton University Press, 1952.

BACH: NUMBERS AND MUSIC

A. Ashton. *Harmonograph: A Visual Guide to the Mathematics of Music.* Walker & Co., 2003.

G. Assayag, H.-G. Feichtinger and F. Rodrigues (editors). *Mathematics and Music: A Diderot Mathematical Forum.* Springer, 2002.

P. Benary. *Musik und Zahl.* Nepomuk, 2001.

H. Bischoff. *Johann Sebastian Bach: The Well-Tempered Clavier.* Warner Bros Publications, 1999.

J. Fauvel, R. Flood and R. Wilson. *Music and Mathematics: From Pythagoras to Fractals.* Oxford University Press, 2003.

R. Haase. *Die harmonikalen Wurzeln der Musik.* Lafite, 1969.

T. Hammel and C. H. Kahn. *Math and Music: Harmonious Connections.* Dale Seymour Publications, 1995.

J. H. Jeans. *Science and Music.* Dover, 1968.

G. Mazzola. *The Topos of Music Geometric Logic of Concepts, Theory, and Performance.* Birkhäuser, 2003.

J. R. Pierce. *The Science of Musical Sound.* W. H. Freeman & Co.,1992.

D. Proust. *L'harmonie des sphères.* Dervy-Livres, 1990.

C. Wolff. *Johann Sebastian Bach: The Learned Musician.* W. W. Norton & Co., 2001.

HOFMANNSTHAL: NUMBERS AND TIME

G. Böhme. *Zeit und Zahl. Studien zur Zeittheorie bei Platon, Aristoteles, Leibniz und Kant.* Klostermann, 1974.

K. G. Denbigh. *Three Concepts of Time.* Springer, 1981.

R. Fischer (editor). "Interdisciplinary perspectives of time." *Annals of the New York Academy of Sciences* 138 (1967), 367–915.

J. T. Fraser. *Time—The Familiar Stranger.* Tempus Books, 1987.

R. M. Gale (editor). *The Philosophy of Time: A Collection of Essays.* Anchor Books, 1967.

J. Götschl. *Erwin Schrödinger's World View: The Dynamics of Knowledge and Reality.* Springer, 1992.

M. Heidegger. *Being and Time.* Harper & Row, 1962.

L. Holford-Strevens. *The History of Time: A Very Short Introduction.* Oxford University Press, 2005.

S. L Macey (editor). *The Encyclopedia of Time.* Samuel L. Macey, 1994.

F. Parise (editor). *The Book of Calendars.* Facts on File, 1982.

C. A. Patrides (editor). *Aspects of Time.* University of Toronto Press, 1976.

C. Sherover (editor). *The Human Experience of Time: The Development of Its Philosophical Meaning.* New York University Press, 1975.

H. Trivers. *The Rhythm of Being: A Study of Temporality.* Philosophical Library, 1985

J. R. Vrooman. *René Descartes.* G. P. Putman's Sons, 1970.

H. Weyl. *Space, Time, Matter.* Dover, 1952.

G. J. Whitrow. *The Natural Philosophy of Time.* Clarendon Press, 21980.

E. Zerubavel. *The Seven-Day Circle: The History and Meaning of the Week.* The Free Press, 1985.

DESCARTES: NUMBERS AND SPACE

C. B. Boyer. *History of Analytic Geometry.* Dover, 2004.

R. Descartes. *Discourse on Method and Meditations on First Philosophy*, third edition. Hackett Publishing Co., 1993.

J. Evans. *The History and Practice of Ancient Astronomy.* Oxford University Press, 1998.

M. Gardner. *On the Wild Side.* Prometheus Books, 1992.

H. Haber. *Space Science: A New Look at the Universe.* Golden Press, 1967.

H. Haber. *Stars, Men and Atoms.* Washington Square Press, 1966.

R. Heath. *Sun, Moon and Earth.* Walker & Co., 2001.

T. L. Heath (translator and editor). *The Thirteen Books of Euclid's Elements.* Dover, 1956.

F. Klein. *Elementary Mathematics from an Advanced Standpoint: Geometry.* Dover, 2004.

M. Minsky. *Society of Mind.* Simon & Schuster, 1988.

E. Moise. *Elementary Geometry from an Advanced Standpoint.* Addison Wesley, 1974

A. S. Posamentier. *Advanced Euclidean Geometry.* Key College, 2002.

C. R. Teed. *The Cellular Cosmogony, or the Earth a Concave Sphere.* Porcupine, 1975.

G. J. Toomer (translator and editor). *Ptolomy's Almagest.* Princeton University Press, 1998.

H. Weyl. *Mind and Nature.* University of Pennsylvania Press, 1934.

LEIBNIZ: NUMBERS AND LOGIC

D. B. Anderson, T. O. Binford, A. J. Thomas, R. W. Weybrauch and V. A. Wilks. *After Leibniz: Discussions on Philosophy and Artificial Intelligence.* Stanford Artificial Intelligence Laboratory Memo AIM-229, 1974.

J. Barwise (editor). *Handbook of Mathematical Logic.* North-Holland, 1977.

E. T. Bell. *The Development of Mathematics.* McGraw-Hill, 1949.

C. B. Boyer. *The History of the Calculus and Its Conceptual Development*. Dover, 1959.

M. Davis. *Computability and Unsolvability*. McGraw-Hill, 1958.

P. J. Davis and R. Hersh. *Descartes' Dream*. Harcourt Brace Jovanovich, 1986.

P. J. Davis and D. Park (editors). *No Way: The Nature of the Impossible*. W. H. Freeman, 1986.

H. Dreyfus. *What Computers Can't Do: A Critique of Artificial Reason*. Harper & Row, 1972.

J. McLeish. *The Story of Numbers: How Mathematics Has Shaped Civilization*. Ballantine Books, 1994.

J. Weizenbaum. *Computer Power and Human Reason: From Judgment to Calculation*. Penguin, 1984.

LAPLACE: NUMBERS AND POLITICS

J. Best. *Damned Lies and Statistics: Untangling Numbers from the Media, Politicians, and Activists*. University of California Press, 2001.

J. Best. *More Damned Lies and Statistics How Numbers Confuse Public Issues*. University of California Press, 2004.

G. R. Boynton. *Mathematical Thinking about Politics*. Longman, 1979.

A. C. Crombie. *Scientific Change*. Heinemann, 1963.

L. J. Daston. *The Reasonable Calculus: Classical Probability Theory, 1650–1840*. Harvard University, 1979.

F. N. David. *Games, Gods and Gambling*. Hafner, 1962.

J. Ford. *How Random Is a Coin Toss?* School of Physics, Georgia Institute of Technology, 1983.

J. W. Forrester. "Counterintuitive Behavior of Social Systems." *Technology Review*, 73(3), 52–68, (1971).

I. Hacking. *The Emergence of Probability*. Cambridge University Press, 1975.

P. L. Houts. *The Myth of Measurability*. Hart Publishing Co., 1977.

D. Huff. *How to Lie with Statistics*. W. W. Norton & Co., 1993.

P.-S. Laplace. *Philosophical Essay on Probabilities*. Springer, 1998.

J. A. Paulos. *A Mathematician Reads the Newspaper*. Anchor, 1996.

E. A. Purcell. *The Crisis of Democratic Theory; Scientific Naturalism and the Problem of Value*. University of Kentucky Press, 1973.

BOHR: NUMBERS AND MATTER

P. A. M. Dirac. *Lectures on Quantum Mechanics*. Dover, 2001.

G. Gamow. *Thirty Years That Shook Physics: The Story of Quantum Theory*. Dover, 1985.

F. Hund. *The History of Quantum Theory*. Barnes & Noble Books, 1974.

A. Messiah. *Quantum Mechanics*. Dover, 2000.

D. R. Murdoch. *Niels Bohr's Philosophy of Physics.* Cambridge University Press, 1989.

A. Pais. *Niels Bohr's Times: In Physics, Philosophy, and Polity.* Oxford University Press, 1994.

A. Pais. *The Genius of Science: A Portrait Gallery.* Oxford University Press, 2000.

K. Strubecker. Einführung in die höhere Mathematik I. Oldenbourg, 1956.

V. F. Weisskopf. *Knowledge and Wonder.* The MIT Press, 1979

E. Wigner. "The Unreasonable Effectiveness of Mathematics in Natural Sciences." *Communications in Pure and Applied Mathematics,* 13(1), 1–14, (1960).

PASCAL: NUMBERS AND SPIRIT

H. Arendt. *The Human Condition.* University of Chicago Press, 1958.

J. L. Borges. *The Library of Babel.* David R. Godine Publisher, 2000.

P. J. Davis and R. Hersh. *The Mathematical Experience.* Birkhäuser, 1981.

A. Huxley. *Literature and Science.* Harper & Row, 1963.

M. Kline. *Mathematical Thought from Ancient to Modern Times.* Oxford University Press, 1972.

R. Musil. *The Confusions of Young Törless.* Translated from the German by Shaun Whiteside, Penguin 2001.

B. Pascal. *Pensees.* Penguin Classics, 1995.

A. S. Posamentier and I. Lehmann. *Pi: A Biography of the World's Most Mysterious Number.* Prometheus Books, 2004.

G. Steiner. *In Bluebeard`s Castle: Some Notes Towards the Redefinition of Culture.* Yale University Press, 1974.

H. Weyl. *Philosophy of Mathematics and Natural Science.* Princeton University Press, 1949.

Index

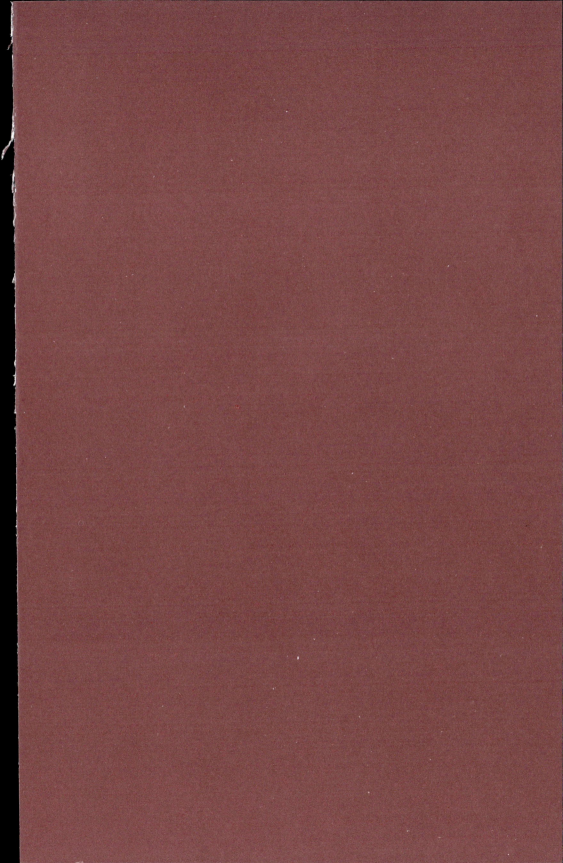